Siemens-Schuckert Aircraft of WWI

A Centennial Perspective on Great War Airplanes

Jack Herris

Great War Aviation Centennial Series #12

This book is dedicated to Dick Bennett, who contributed so much to it, including his knowledge, shipment logs, fighter summary, photographs, and his exceptional drawings of the SSW D.III and D.IV. In turn Dick wants to acknowledge the late historian Peter M. Grosz for his generosity in providing much of the original material.

Acknowledgements

My sincere thanks to Colin Owers, Aaron Weaver, Reinhard Zankl, Lance Bronnenkant, Adam Wait, and Richard Andrews for photographs, and Greg VanWyngarden for photographs and helpful suggestions and material. Special thanks to Dick Bennett for information, photos, and drawings, Bob Pearson for his color profiles, Aaron Weaver for his digital photo editing and cover design, Dick Bennett, Martin Digmayer, and Andreas Horn for scale drawings, and the Deutsches Technikmuseum in Berlin and the Museum of Flight in Seattle for photographs. Thanks again to Andreas Horn for his information on the Siemens-Schuckert airship. Any errors are my responsibility.

The cover painting by Steve Anderson illustrates three SSW D.III aircraft of *Kest* 5. These aircraft and another from *Kest* 5 landed in Switzerland on Nov. 13, 1918. Please see Steve's website at: www.anderson-art.com

Color aircraft profiles © Bob Pearson. Purchase his CD of WWI aircraft profiles for $50 US/Canadian, 40 €, or £30, airmail postage included, via Paypal to Bob at: bpearson@kaien.net

For our aviation books in print and electronic format, please see our website at: www.aeronautbooks.com. I am looking for photographs of the less well-known German aircraft of WWI to illustrate future titles. To help with photographs please contact me at: jherris@verizon.net.

Interested in WWI aviation? Join The League of WWI Aviation Historians (**www.overthefront.com**) and Cross & Cockade International (**www.crossandcockade.com**).

www.aeronautbooks.com

Table of Contents

Above: The Siemens-Schuckert airship SSL (most probably the first version, "*SSL 1*") in its revolving shed at Biesdorf near Berlin. There was only one Siemens-Schuckert airship designed by the engineers Dietzius and Haas under the director of the "*Kriegs- und Schiffsbautechnische Abteilung*" Otto Krell. Krell wanted to build a non-rigid airship with the dimensions of a contemporary Zeppelin airship (*LZ 3*). Unofficially designated *SSL 1*, it made its maiden flight on January 23, 1911. Built for the army, it was modified several times. After 28 flights a major modification was a larger envelope to increase volume from 13,500 to 15,000 m³. The engine gondolas remained the same but more powerful engines were installed and the mid-ship gondola was lengthened. A new rudder and elevator were installed so some people gave this enlarged version the (unofficial) designation *SSL 2*, but the Siemens people always talked about the *SSL*. This large (120 m), innovative airship did not meet Army requirements, and after another 36 successful flights, it was bought by the army in March 1912 but was dismantled without making any more flights.

Above: Airmen of postwar *Flieger-Abteilung* 424 in front of an SSW D.IV flown by the unit.

Above: *Lt.* Karl-August von Schoenbeck (right) in front of an SSW D.IV of postwar *Flieger-Abteilung* 424.

Introduction

Above: SSW D.IV 7555/17, the true prototype of the D.IV, made its first flight June 18, 1918. It was delivered to Adlershof in August as the D.IV static test airframe. The D.IV epitomizes the SSW contribution during the war. Powered by an innovative counter-rotary engine produced by Siemens-Halske, the sister company of Siemens-Schuckert Werke, the D.IV demonstrated exceptional climb, ceiling, and maneuverability and proved itself in combat. Most pilots who flew the SSW D.III and D.IV believed them superior to any other operational fighter, Allied or German, of the war.

Siemens was founded by Werner von Siemens on October 12, 1847 as Telegraphen-Bauanstalt von Siemens-Halske, a private company located in Berlin. It grew and was incorporated as a public company in 1897. Siemens & Halske acquired a new site northwest of Berlin in 1897 and gradually concentrated all operations there; by 1914 Siemensstadt, an all-new community, was created. In March 1903 the heavy-current divisions of Siemens & Halske were merged with the Schuckert & Co. to form Siemens-Schuckertwerke GmbH. By 1907 Siemens (Siemens & Halske and Siemens-Schuckert) had 34,324 employees, making it the seventh-largest company in the German empire by number of employees. Among manufacturing companies it was second only to Krupp.[1] In 1966 SSW was reorganized as Siemens AG, which exists today as Europe's largest engineering company.[2]

SSW's first foray into aviation was construction of the SSL non-rigid airship that first flew in 1911. Despite satisfactory flight performance, it was not robust enough to meet military requirements and SSW abandoned airship construction.

In 1909 SSW established an airplane department to build Burkhardt's designs, but these were mediocre and the department became dormant. However, in 1914 the department was reactivated due to an urgent request for aircraft to supply the German Army. Under the leadership of Professor Dr. Walter Reichel, a director of the Siemens-Schuckert Dynamowerke, SSW built the first R-plane to reach the front and also developed and built the exceptional D.III and D.IV fighters of 1918. SSW also built nearly 100 Albatros C.IIIc(SSW) and 80 Gotha G.IV(SSW) aircraft under license.

In addition to SSW building approximately 600 aircraft, including types under license, its sibling Siemens-Halske company designed and built a series of innovative rotary engines. The Sh.III counter-rotary engine was used in a number of designs and was largely responsible for the exceptional climb performance of the fighters that used it.

Not content with these achievements, SSW also designed and tested a series of air-launched guided missiles! Although these did not achieve operational status, they were of major technical and historical significance and are therefore briefly documented here along with the Siemens-Halske engine series.

Above: SSW D.I airframes 3524/16, 3525/16, 3534/16, and others being painted at the SSW factory at Nürnberg in 1916. Aircraft in the foreground have the four-louver engine panel while those in the background have the later panel with three louvers. Camouflage is the three-color scheme. The interior views show the factory was clean and modern for 1916.

Facing Page Top Left: An Albatros C.IIIc(SSW) in front of the SSW factory (logo near peak of roof) at Nürnberg in 1917. When production of the SSW D.I was cancelled, SSW was left with unused production capacity and in late 1917 was assigned to produce the Albatros C.IIIc under license as the Albatros C.IIIc(SSW). One hundred aircaft, 14000–14099/17 were ordered and 93 had been delivered by December 1918 when production stopped. Similarly, a contract for 100 LVG B.III(SSW) trainers, 2300–2399/18, was awarded to SSW in August 1918, and two were under construction at the Armistice.

Facing Page Bottom Left: SSW D.I fuselages being built at the SSW factory at Nürnberg in 1916.

Frontbeststand Inventory of All SSW Fighters at the Front

Manufacturer and Type		1914 31 Aug	31 Oct	31 Dec	1915 28 Feb	30 Apr	30 Jun	31 Aug	31 Oct	31 Dec	1916 28 Feb	30 Apr	30 Jun	31 Aug	31 Oct	31 Dec	1917 28 Feb	30 Apr	30 Jun	31 Aug	31 Oct	31 Dec	1918 28 Feb	30 Apr	30 Jun	31 Aug
SSW	E.I													5	2	1	1									
SSW	D.I																	2								
	D.III																							4	4	6
	D.IV																									3

Above: This *Frontbestand* inventory of all SSW fighters at the front during the war reveals that their numbers were too few to make a major impact. The SSW E-types arrived late on the scene and were obsolete when they reached the front; consequently, few were ordered and fewer still saw operational service. Although 95 SSW D.I fighters were built, most were used for training because they too were obsolescent when built. The excellent SSW D.III and D.IV were built in larger numbers but constrained engine availability limited production and front-line service. Although they were fully competitive fighters, and its pilots thought the D.IV the best fighter of the war, too few were available to make a significant impact. SSW fighter numbers grew significantly in the war's final months but no official data is available for the end of October.

Above: The SSW D.IIc *kurz*, 7550/17, undergoing static load testing. The airframe is inverted and the engine is mounted.

Below: The SSW D.IIc *kurz*, 7550/17, undergoing static load testing. The airframe is upright and no engine is mounted.

Above: The SSW factory producing D.III fighters and Gotha G.IV(SSW) bombers; 80 Gothas were built under license.
Below: The SSW D.IIe, 7553/17, undergoing static load testing. The airframe is inverted and no engine is mounted.

Above: SSW and *Idflieg* officials tour the factory where SSW D.III fighters are under construction.

Below: Cockpit of a late series SSW D.III under construction at the factory.

Facing Page Above: SSW D.IV fuselage being built in the construction jigs.

Facing Page Below: Cockpit of an SSW D.IV; the white rectangle at center right is a rigging diagram.

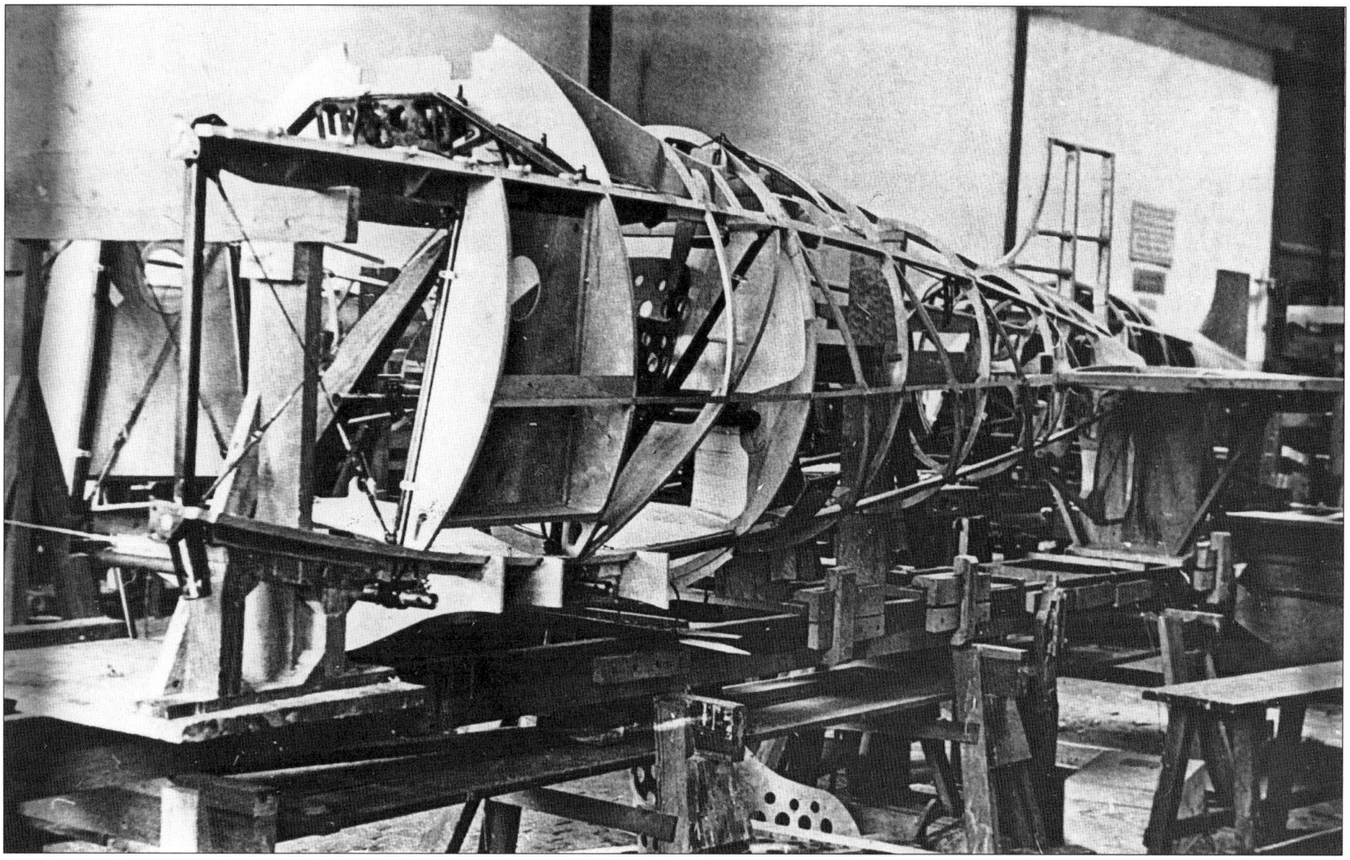

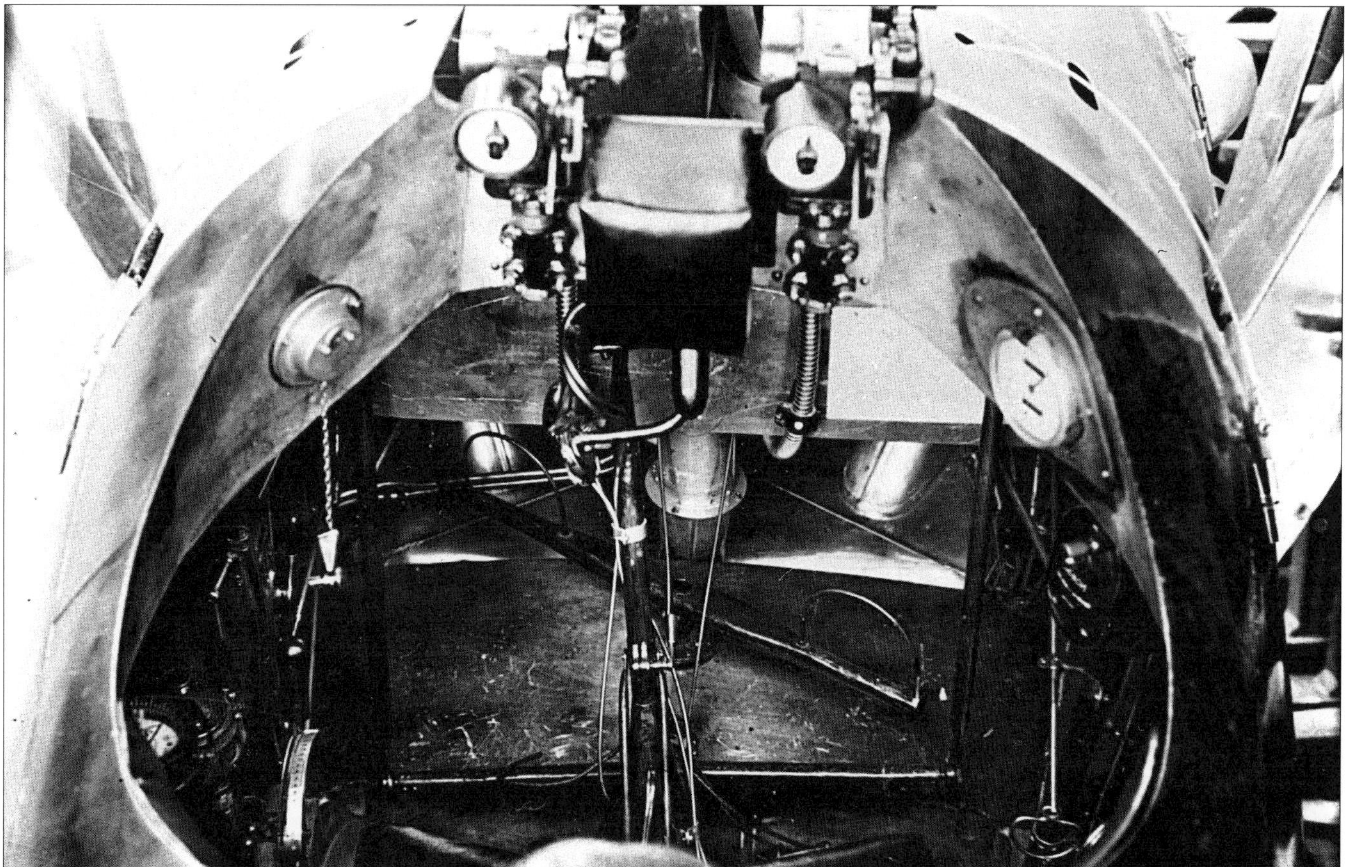

Pre-War & Early War SSW Aircraft

The first SSW airplane was a biplane based on the Wright Flyer. Powered by a 50 hp Argus engine, it was built in 1909 and first flew on December 31, 1909. Damaged on landing after this flight, it was modified and flew again on March 9, 1910 carrying its designer, Bourcart, and one passenger. On March 11 it flew for a kilometer with three people aboard, but crashed when caught in a cross-wind. The next SSW design was a high-wing monoplane first flown in May 1911. A primitive design, it was not successful. A second monoplane to the Bleriot configuration was completed in September 1911, but its failure caused SSW to temporarily abandon aviation.

SSW "Bulldog"

Shortly before the war Swedish designer Villehad Forssman, who had an office in Berlin, designed a monoplane known as the Bulldog for Prince Friedrich Sigismund of Prussia. Soon after the war started, Forssman offered SSW his design. The War Ministry desperately needed warplanes, and with their support SSW hired Forssman to supervise construction of two of his Bulldogs for reconnaissance. SSW began assembly of the aircraft at their Dynamo Works in their community of Siemensstadt.

The first Bulldog, built in late 1914, was powered by a 100 hp Mercedes D.I in-line, six-cylinder engine. The second Bulldog, powered by one of the first 110 hp Siemens-Halske Sh.I counter-rotary engines, first flew on April 18, 1915. Both featured side-by-side seating for two crewmen. Both Bulldogs were rejected by the Army for mediocre flying qualities and poor performance. After modification, the rotary-powered Bulldog was used for air-launching trials for models of the missiles SSW was developing.

Above & Left: Two views of the "Bulldog" monoplane Forssman designed and built for Prince Friedrich Sigismund of Prussia. The royal crest was applied to the rudder. Engine was a Mercedes.

Above: Front quarter view of the 100 hp Mercedes D.I-powered "Bulldog" built by SSW to Forssman's pre-war design. Seating was two crewmen side-by-side. No armament was fitted. The radiators are fitted between the wing-roots and the wheels.

Right: The Bulldog powered by a 110 hp Sh.I counter-rotary. Its mediocre handling qualities were no better than the Mercedes-powered version.

Above: Forssman's 100 hp Mercedes D.I-powered "Bulldog" with Forssman standing in front of it; the vertical tail has been enlarged from the aircraft for Prince Sigismund. This aircraft and the other "Bulldog" powered by a 110 hp Siemens-Halske Sh.I counter-rotary engine were both rejected by the Army for poor performance and mediocre flying qualities.

SSW B.I

In the summer of 1915 brothers Bruno and Franz Steffen, both famous pre-war pilots and aircraft builders who joined SSW in December 1914, began designing single-engine military aircraft. The B.I was a two-seat reconnaissance aircraft powered by a 110 hp Siemens-Halske Sh.I and was the only one of these designs that was not a fighter. The B.I two-bay biplane achieved 95 mph, a good speed for the time, and was sent to *BAO* (*Brieftauben Abteilung Ostende* – Carrier Pigeon Section Ostend), a code name for one of Germany's original bombing units, at the request of its commander, *Freiherr* von Thuna. The only B.I built, it crashed in November 1915 and was not repaired.

Above & Left: Two views of the SSW B.I reconnaissance biplane. It was a neat, two-bay design for two crewmen. Engine was a 110 hp Sh.I counter-rotary. No armament was fitted.

Above & Below: These two views of the SSW B.I reconnaissance biplane reveal that the fuselage and tail were closely related to those of the SSW E.I monoplane fighter. The only B.I built was sent to *BAO* at the request of its commander, where it crashed in November 1915.

Left: This view of the SSW B.I emphasizes its 9-cylinder 110 hp Sh.I counter-rotary.

Right & Below: Two views of the SSW B.I. The 110 hp Sh.I counter-rotary required a front bearing support. The Siemens-Schuckert company logo is on the hangar doors below.

SSW Bombers

In October 1914 SSW entered the giant bomber business with two designs, one by Forssman and the other by the Steffen brothers. These two designs were unrelated and dissimilar, but both were influenced by the success of the pre-war Sikorsky *Grand*, the world's first successful four-engine airplane. The *Grand* flew first as a two-engine aircraft on March 2, 1913 and as a four-engine aircraft on May 13, 1913. Initially the engines were in tandem, but in June the rear engines were moved to the front of the lower wing. In this final configuration the *Grand* was renamed *The Russian Knight* and became the development aircraft for the *Il'ya Muromets* four-engine transport, the first prototype of which flew on December 11, 1913. In April 1914 the Russian Navy ordered a float-equipped version as a long-range reconnaissance aircraft, and the first prototype was converted to floats and delivered to the Navy in late May. The second *Il'ya Muromets* made a record-breaking long-distance round-trip flight in early July. After war started, the second *Il'ya Muromets* and another were delivered to the Russian Army on August 31, 1914, and five more aircraft were delivered by December, giving the Russian Army a long-range reconnaissance bomber. At that time no other nation had a comparable warplane.

Development of the Forssman "R" was terminated after the construction and traumatic testing of one aircraft. SSW built seven aircraft, R.I through R.VII, to the Steffen design and built R.VIII, the world's largest airplane at the time, to a more advanced design. During 1917 and 1918 SSW also built 80 Gotha G.IV bombers under license in two orders of 40 each; these serials were 1055–1094/16 and 200–239/17.

SSW Forssman "R"

Construction of the Forssman "R", SSW's first experience with R-plane building, began in October 1914. Its design was clearly inspired by the Sikorsky *Il'ya Muromets* and construction was completed in the first half of 1915. The initial configuration had four uncowled, 110 hp Mercedes D.I engines independently mounted on the lower wing, each driving its own propeller. The cabin was fully glazed for an excellent field of view and, like its Sikorsky inspiration, projected just a small distance

Below: The Forssman "R" under construction clearly shows the influence of the Sikorsky Il'ya Muromets on its design.

Above: Forssman "R" in original configuration under construction in the SSW factory. The blunt nose is essentially in line with the leading edge of the wings.

Below: Forssman "R" in its second configuration after an observer's/gunner's 'pulpit' was added to the nose in an attempt to compensate for the extreme tail-heaviness evident during the first attempted flights.

Above & Below: The Forssman "R" in its second configuration after addition of the observer's/gunner's pulpit to the nose. The engines were four 110 hp Mercedes D.I's with individual propellers. Originally all engines were un-cowled, but the inboard engines now have cowlings. These views show the middle bay of bracing had only a forward strut; the struts that would normally brace the aft wing spars were not fitted. At this stage the aircraft still had a single rudder.

Above & Below: The Forssman "R" in its third configuration after being modified to have two rudders instead of the inadequate single rudder. All engines now have cowlings and the 'missing' rear, middle interplane struts were added.

in front of the wings.

Early test hops quickly revealed the design was seriously tail heavy, and an observer's/gunner's pulpit was added in front of the pilot to add weight to the nose. The biplane wing cellule had three bays, but the aft strut was omitted from the middle bay, probably in an attempt to save weight. This weakened the cellule and the 'missing' strut had to be installed. To improve directional control and stability a second rudder had to be added. Although the wing struts and landing gear were too strong and heavy, overall the structure was weak. In addition to poor handling qualities and a weak structure, the Forssman R was also under-powered. The design was a failure and SSW and Forssman parted ways.

This should have been the end of the story.

Unfortunately, SSW was reluctant to accept failure and tried to salvage the design, resulting in a painful, protracted development process that was foredoomed to failure. To improve performance the inboard engines were replaced with 220 hp Mercedes D.IV engines and the outboard engines were cowled and moved to a higher position thought to improve propeller efficiency. In addition, the nose was rebuilt to a more streamlined form. In its modified form the machine was tested by *Lt.* Walter Höhndorf (later to become a fighter ace awarded the *Pour le Mérite*), but the Forssman over-turned on landing.

Despite the resulting damage, SSW again rebuilt the aircraft, adding a more rounded nose. By now the Forssman had acquired the reputation of being a white elephant and no pilot wanted to fly it.

This Page: The Forssman "R" in its fourth configuration after crashing in September 1915 during landing while being tested by *Lt*. Höhndorf, who later became a fighter ace and was awarded the *Pour le Mérite*. By this time the inner engines had been replaced with 220 hp Mercedes D.IV engines for more power. The Forssman was rebuilt after this accident.

Above: The Forssman "R" in its fourth configuration before the crash.

Above & Below: The Forssman "R" in its fifth configuration after being rebuilt from its September 1915 crash. The nose has been extensively rebuilt with large windows and a gunner's position. All the engines are neatly cowled with spinners; the inboard engines are the 220 hp Mercedes D.IV and the outboard engines are the 110 hp Mercedes D.I. Its neat appearance belies the airframe's structural weakness and poor aerodynamics.

Above & Below: The Forssman "R" in its sixth and final configuration. The nose has again been extensively rebuilt. In this configuration *Idflieg* finally accepted the Forssman based on reduced requirements, but further development was abandoned.

Anxious to sell the aircraft to the Army, SSW offered the Steffen brothers 10% of the delivery price if they would perform the acceptance flight. *Idflieg* agreed to reduced specifications in return for a lower price. The Steffen brothers examined the aircraft and determined it was flyable.

In October 1915 Bruno Steffen flew the aircraft on its acceptance flight. He flew alone; all of the people he invited to accompany him declined, including members of *Idflieg's* acceptance commission! The flight was an adventure. The aircraft was very tail heavy and Bruno struggled to control it. After exceeding the required 2,000 meter height in 28 minutes, Bruno was descending when one of the outboard engines failed due to inadequate fuel pressure. While trying to restart the engine by manually pumping fuel, all the other engines stopped! Despite total engine failure, Bruno was able to land the aircraft safely and it was accepted by *Idflieg* in April 1916 for training. However, further development was cancelled.

While running the engines on the ground shortly after acceptance, the fuselage broke in two just aft of the wings due to excessive vibration and the aircraft was scrapped. A relieved Bruno Steffen was glad no one had been killed in the aircraft. Despite extensive development, being modified at least four times, the Forssman was a complete failure.

Above: Forssman "R" in its final configuration at Johannisthal.

Below: Forssman "R" in its final configuration ready for take-off at Johannisthal.

Above: Forssman "R" in final configuration during a take off at Johannisthal.

Above & Right: Forssman "R" in final configuration after breaking its back during engine run-up on the ground at Johannisthal. Fortunately, no one was injured and the aircraft was scrapped after this debacle.

SSW R.I

In 1908 the Steffen brothers, Bruno and Franz, founded the Steffen aircraft works in Neumünster (near Kiel). After first building one airship, they built a series of airplanes, and in one of those, the *Falke*, Bruno Steffen broke the German endurance record in 1913. Additionally, they also operated a successful flying school. Despite their relative aviation success, they were called to serve in the army upon outbreak of war. Moreover, lack of capital also precluded further independent development, and in December 1914, after release from their regular military duties, the brothers joined SSW. The Steffens immediately started designing the SSW R.I, which made its first flight in May 1915.

The brothers decided that a design with centralized engines offered several important advantages. Concentrating the weight near the center of gravity would make the aircraft more maneuverable and, most importantly, the engines would be accessible to on-board mechanics during flight should repairs be needed. They also expected drag would be reduced, propeller efficiency would be improved, and the ability to shut down an engine in flight would extend range and endurance.

Another critical factor at the time was lack of ability to feather a propeller. To feather a propeller means to turn the propeller blades of an inoperative engine into the wind to stop rotation and reduce the drag as much as possible. This requires a controllable propeller having a pitch-change mechanism to rotate the propeller blades into the wind upon engine failure. However, such a propeller was not available during WWI; the propellers available were all fixed pitch. Upon engine failure, a fixed-pitch propeller continues to rotate as the aircraft's speed pushes the propeller forward, and such a 'wind-milling' propeller creates drag approximating that of a parachute of the same diameter. The result was that twin-engine aircraft and some multi-engine aircraft could not maintain flight after engine failure and a wind-milling propeller. Without a controllable propeller that could be feathered upon engine failure, centralizing the engines and using a clutch to decouple a failed engine from the propellers was one way to avoid a wind-milling propeller in case of engine failure. For this reason many R-planes were designed with central engine arrangements, and the SSW R.I was the first of these.

Originally designated SSW G.31/15, (G for *Grossflugzeug* [large aircraft]), the designation was changed to SSW G.I 32/15 and finally to SSW R.I 1/15 on November 6, 1915. The SSW R.I was a distinctive biplane with three 150 hp Benz Bz.III engines grouped together in the center fuselage. The

Above: The SSW R.I 1/15 was powered by three 150 hp Benz Bz.III engines mounted in the fuselage and all connected to a central transmission and drive system. The two propellers were powered via drive shafts. The aircraft's type designation was SSW R.I and its military serial number was 1/15; only one aircraft was built.

Above: The SSW R.I at Neumünster in May 1915. From the left, the people are Karl Friedrich von Siemens, Franz Steffen and his wife, *Dr.* Walter Reichel, Bruno Steffen, and *Dipl. Ing.* Dinslage.

engines were connected to a common gearbox via clutches. Two engines were mounted side by side in front of the gearbox on steel engine bearers that were part of the airframe, while the third engine was mounted behind and below it. Unfortunately, the clutch system of the R.I did not enable the mechanic to disengage a failed engine. The gearbox was reliable but there were issues with the clutch and transmission system. Three radiators were fitted around the nose.

The crew included a pilot, an aircraft commander acting as reserve pilot, two gunners, and one mechanic. The fuselage was unique; two triangular tail booms supported the rudder and tailplane. This provided a good field of fire to the rear, but the position was covered with fabric in the R.I. The pilot's cabin was open in front because the pilots insisted on feeling the wind in flight to help them control the airplane. This is not as odd as might be supposed given the limited flight instrumentation of the time; pilots depended on their senses, including the feel of the wind, the sound of the engines and the wind through the wires, etc., for important feedback on aircraft performance.

An upper gun position was located behind the pilots' seats, which were themselves located above the gearbox. This position was an enclosed platform with an access hatch. A ventral gun position was located between the robust landing gear assemblies;

Above: The SSW R.I remained in active training service into 1918. Here it is seen at Döberitz.

the gunner was prone and fired downward toward the rear. In addition to these machine gun positions, the aircraft commander and mechanic were given repeating carbines to help defend against attack. These were not very effective and were most useful as a morale booster.

The three-bay wings were built of wood and the fuselage was built of steel tubing; both were fabric-covered. Some aluminum panels partially covered the nose. Armor plating weighing 200 kg was originally specified but the armor was eliminated by *Idflieg*. To reduce the pilot's control forces the pilot controlled small auxiliary ailerons that in turn actuated the ailerons mounted on the upper wing. The rudder and elevators were similarly assisted by servo controls.

The R.I was built at the SSW Dynamowerk in Berlin, but the aircraft was disassembled and moved to the Steffens' facility in Neumünster for flight trials for secrecy. Piloted by the Steffen brothers, the R.I first flew on May 24, 1915, only five months after starting the design. The R.I proved to be stable and easily maneuverable, according to one pilot not unlike a typical two-seater.

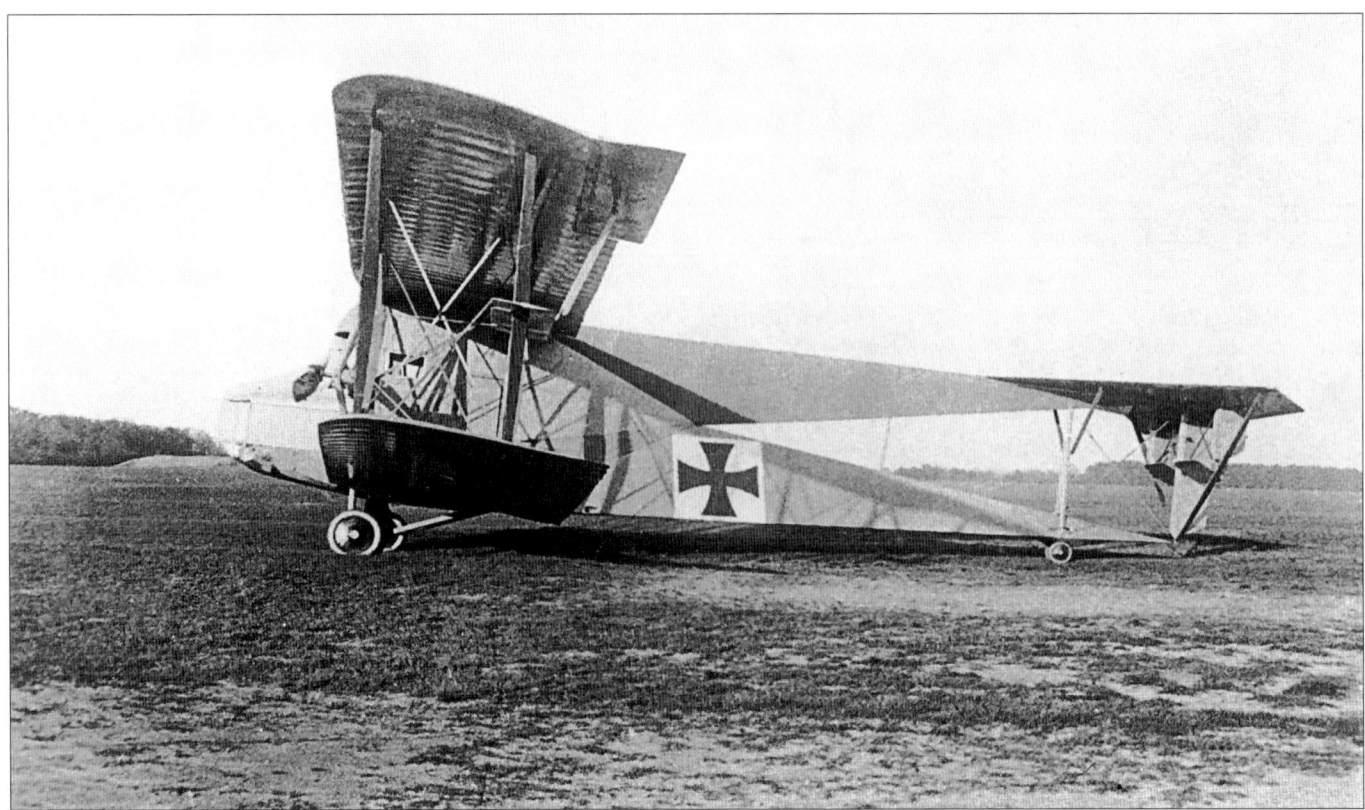

Above: This side view emphasizes the unusual fuselage design of the SSW R.I. This basic design carried through the SSW R.II to R.VII.

The first flight was terminated early due to excessive gearbox oil temperature. Installation of forced oil circulation and switching to a lighter grade of oil remedied that problem. Failures of the u-joints in the propeller drive shafts occurred due to vibration; these were remedied by adding stiffener tubes fore and aft.

The R.I had a maximum level speed of 128 km/hr with full load and could maintain level flight on two engines with the third at idle. With only one engine running the R.I could glide 50 km starting from an altitude of 2,000 meters.

In June 1915 the R.I passed its acceptance flight while carrying seven passengers. Before final delivery to the *Fliegertruppe* on July 26 an additional 24 training and orientation flights were completed. During one of them, with *Lt*. Walter Höhndorf piloting, an engine problem forced an emergency landing, but the problem was quickly repaired and the flight was continued the same evening.

During a training flight in early August all three engines failed one after the other on take-off at only 8 meters altitude, resulting in a crash landing and severe damage to the R.I. Fortunately, the crewmen escaped injury. The cause was determined to be foreign matter in the fuel tanks that blocked fuel flow to the engines. The factory repaired the R.I and

it flew again in late September, at which time it was sent to the Eastern Front for operational evaluation. At Warsaw the R.I had to be parked in the open in heavy rain. The wings absorbed gallons of rainwater and the R.I would not take off the next morning until the water was drained. Continuing on, the starboard transmission coupling broke and the R.I had to be sent back to Döberitz for repair on October 15. The R.I was returned to the Eastern Front but no bombing raids were undertaken due to continuing unreliability. It was feared the aircraft would be forced down in enemy territory. In March 1916 the R.I was dismantled and shipped back to the SSW factory in Berlin. Unfortunately, due to clearance problems it was damaged during transport.

In June 1916 SSW completed repairs to the R.I, after which it was assigned to train R-plane crews at Döberitz. A year-end report of 1917 by the *REA* (*Riesenflugzeug Ersatz Abteilung* – giant aircraft replacement unit) described the R.I as larger and heavier than before, possibly due to modifications during its factory repairs.

During 1917 the R.I performed 97 training flights and 26 orientation flights, an excellent record for this complex, early R-plane that was still in training service in late February 1918.

Above: Rear view of the SSW R.I.

SSW R.II – R.VII

Based on the successful launch of the SSW R.I, on June 10, 1915, *Idflieg* gave SSW a contract to build six more R-planes. These were initially designated G.32–37/15, then changed to G.33–38/15, and finally to R.II 2/15 to R.VII 7/15. The specification required a maximum speed of 135 km/hr (a small increase over the R.I's speed), climb to 2,000 meters in 35 minutes and 3,000 meters in 70 minutes carrying a useful load of 2,450 kg, a crew of five like the R.I, and fuel for an endurance of six hours. Specified armament was a cannon in the forward turret and two machine guns, one in the dorsal position and another in the ventral position, plus 600 kg of bombs. However, the cannon was never fitted and a lighter mount for a machine gun was installed. Armor to protect the pilot was initially specified, but *Idflieg* later dropped this requirement to save weight.

Idflieg required the aircraft to maintain level flight on the power of two engines, and wanted, but did not require, the aircraft to be able to take off on the power of two engines with full crew and armament (less bombs) and one hour of fuel.

Idflieg also specified a strict aircraft delivery schedule. The first was to be delivered by the end of September, 1915, the second by the end of October, and the remaining four aircraft at eight-day intervals thereafter. The deadline for final delivery of all aircraft in the order was April 1, 1916.

These SSW R-planes were designed to use three 240 hp Maybach HS engines, the most powerful aero-engines then available. Originally designed for airship use, the Maybachs proved unable to withstand the more rigorous demands of aircraft use, especially prolonged running at maximum power during take-off and lengthy climbs to altitude. Engine failures were depressingly common and greatly impeded flight testing. In fact, the engine's reliability was marginal even in airships. Finally the aircraft had to be re-designed to use either the 200 hp Benz Bz.IV or the 260 hp Mercedes D.IVa six-cylinder engines. This was a major problem because the central engine location was integral with the structure, and the engine bearers, clutch, and central transmission system all had to be re-designed for the new engines, causing a great deal of additional work. Overall, the Maybach engine debacle delayed delivery of these aircraft by nearly two years.

The lower-power Benz engines were installed in R.III, R.IV, and R.V. Flight tests soon revealed that these aircraft were under-powered and could not meet *Idflieg's* climb and useful load requirements. Wing span and area of these aircraft had to be enlarged for more lift to compensate for the lower power of the Benz engines.

The more powerful Mercedes engines were installed in R.II and R.VII. However, the greater weight of the Mercedes engines again meant the wing span and area had to be enlarged, this time to compensate for the weight increase.

Above: The SSW R.II 2/15 under construction at the SSW Dynamowerk in October, 1915.

Below: Part of the SSW work team posses in front of the SSW R.II 2/15 at the SSW Dynamowerk.

Above: This side view of SSW R.II after being rebuilt with longer wings emphasizes the unusual fuselage design of the SSW R-planes. This basic design carried through the SSW R.II to R.VII.

Above: SSW R.II with its original wings. The nose radiator installation is much cleaner than on the R.I. Despite the improved cooling, the experimental 240 hp Maybach HS engines continually over-heated and were subject to many other failures. The Maybach engines were the Achilles heel of the SSW R.II–R.VII.

Above: This front view of the rebuilt SSW R.II emphasizes the long wing-span after addition of the "supplementary bays" required for greater lift.

These aircraft, originally designed with three bays of struts, had to have wings enlarged to four, five, or even six bays due to the increased span needed for greater wing area. These "supplementary bays" were installed between the wing center section and the outboard wing sections. The variety of engines, wing variations, and other changes during development meant that no two of these aircraft were alike.

To facilitate in-flight engine servicing and repairs, the engine room was enlarged and each engine was fitted with a separate clutch to enable a failed engine to be de-coupled and stopped while the running engines continued to power the propellers. If the mechanic could repair the engine in flight, it could be restarted and re-engaged with the clutch to power the transmission and propellers.

However, despite the enlarged engine room, the space was still very cramped, impeding the mechanic's ability to service the engines in flight. In flight the engine room was hot and often filled with hot, toxic gases from exhaust leaks. The mechanic had to avoid contact with hot exhaust manifolds, which was not easy during flight. The forward engines were so close together that the cylinder head of the left engine had to be removed to provide enough room to replace the exhaust gaskets on the right engine!

Another problem with the complex, centralized power system with internal engines was that any engine or transmission repairs were hardly possible without practically tearing the whole fuselage apart. Additionally, misalignment of assemblies caused frequent drive failures. Lengthy, painstaking development work eventually resolved most of these problems, resulting in reasonable reliability. The SSW R-planes eventually flew a significant number of successful day and night missions. However, the delays meant that, by the time the SSW R-planes arrived at the front, more modern Zeppelin-Staaken R-planes with decentralized power were also in service and gave greater reliability with less development effort. On the positive side, the SSW R-planes had excellent flying characteristics.

R.II 2/15

This was the first aircraft of this batch to be completed and flown, its first flight being on October 26, 1915. However, R.II used the ill-fated Maybach engines, and test-flight experience soon required their replacement. After discussion, *Idflieg* decided on installation of 260 hp Mercedes D.IVa engines, but by then a shortage of skilled workmen caused SSW to focus on completion of R.V–R.VII and R.II was temporarily stored until the workforce could get to it.

Work finally restarted in early 1917. By then SSW had experience with the Mercedes-powered R.VII and the R.II was completely rebuilt with extended-span wings and enlarged tail surfaces. The R.II was finally delivered to *REA* on June 29, 1917, but by that time it was not able to compete with the new Staaken R-planes and was relegated to training at the R-plane school at Döberitz. In 1918 it was transferred to *REA* Cologne and later crashed.

Above: The SSW R.III in flight in its original configuration.

Right: The SSW R.II 2/15 in final configuration with extended-span wings.

Below: The SSW R.II; the nose gun position was intended for a cannon that was never fitted.

Above: The SSW R.IV as completed with its original, short-span wings. Originally accepted with the problematic Maybach HS engines, after serious problems the R.IV was re-engined with three 200 hp Benz Bz.IV engines.

R.III 3/15

R.III was delivered about a month after R.II. However, R.III was also sabotaged by its Maybach engines and crashed in early 1916 due to their failure. At *Idflieg's* direction, R.III was stored until SSW had the resources to rebuild it. Later in 1916 R.III was rebuilt with enlarged wing span, 200 hp Benz Bz.IV engines, a reinforced tail, and numerous detail improvements. *Idflieg* realized the modified R.III was not suitable for combat use and R.III was delivered to *REA* for training on December 12, 1916. It was still in training service in late February 1918.

R.IV 4/15

R.IV was singled out for extensive development of the Maybach HS engines and related modifications to prove these engines were flight-worthy, and as a result may have been the only airplane powered by these engines to be accepted by *Idflieg*. The R.IV was delivered on January 29, 1916, and immediately experienced severe engine problems, not performing a useful flight until May. Not being responsible for the engine problems, SSW asked *Idflieg* to accept R.IV and continue the test program. *Idflieg* agreed and accepted R.IV on August 27, 1916, after a flight was completed to lowered requirements. By this time the R.IV wing span had been enlarged.

The R.IV was repaired in November, perhaps as a result of a crash. At this time the Maybach engines were replaced by Benz Bz.IV engines and the wing span was again increased. The first flight of the rebuilt aircraft was on March 14, 1917. R.IV joined *Rfa* 501 at Vilna on the Eastern Front on 27 April and flew several operational missions. The R.IV was then used as a trainer at Vilna and later returned to Berlin and made a safe emergency landing near Spandau in June 1918. By the end of August it was again repaired.

R.V 5/15

R.V was nearly complete when it was decided to replace the Maybach HS engines with 200 hp Benz Bz.IV engines, requiring a complete rebuild. R.V was delivered to *REA* on August 13, 1916 and then flown to *Rfa* 501 at Vilna. This delivery flight was a true adventure characterized by clutch and engine failure. Repair required complete disassembly of the gearbox to reach a failed bearing.

R.V flew a number of bombing missions from late 1916 into early 1917. However, in mid-February R.V was severely damaged in a night landing accident. The remains were shipped back to Döberitz for use as spare parts for *REA*.

R.VI 6/15

R.VI was partially complete when it was decided to replace the Maybach HS engines with 200 hp Benz Bz.IV engines, again requiring a complete

Above: The SSW R.IV 4/15 shared the typical SSW R-plane profile. The servo-tabs between the ailerons that reduced the pilot's aileron control forces are clearly shown.

Below: The SSW R.V as completed with its original, short-span wings.

Above: The rebuilt SSW R.V 5/15 was powered by 200 hp Benz Bz.IV engines. It arrived at the *Rfa* 501 airfield at Vilna on the Eastern Front on September 4, 1916.

rebuild. This included an increased wing span. It was shipped from the factory on April 25, 1916. Several successful test flights were made at the factory, including one that established an unofficial world record of six hours in the air carrying a useful load of 2,400 kg. However, the secrecy required by the war prevented any publicity.

Despite this achievement, the climb and ceiling of the R.VI were marginal and the delivery requirements had to be relaxed to a climb to 3,000 meters in 90 minutes with a reduced useful load of 1,410 kg. R.VI passed these requirements, as R.III and R.V did later, and was delivered to *REA* on July 20, 1916, from which it was flown to *Rfa* 501 on August 7. R.VI performed is first operational bombing mission on September 3 but suffered an operational career checkered by mechanical problems and did not complete many operational missions. R.VI was dismantled in November 1917 as being of no further use.

R.VII 7/15

R.VII was the first aircraft of the batch to receive the 260 hp Mercedes D.IVa engines, a change that required extensive rebuilding to strengthen the airframe and power assembly for the larger, more powerful engines. In fact, completely new, larger, stronger wings had to be built and the tailplane had to be enlarged to compensate.

The R.VII was first test-flown on January 15, 1917. The wings were determined to be too heavy and during its test program new, lighter wings were fitted. R.VII was delivered to *REA* on February 11 and flown to *Rfa* 501 at Vilna on the 26th of that month. At Vilna a number of field modifications were made, including painting the wing undersurfaces gray for night operations and adding bomb racks for additional bombs, bringing its total bomb-load to 750 kg. R.VII made its first flight from Vilna on March 12, 1917 and its first operational flight on the 15th. However, failure of the spur gear of the left engine during take-off resulted in an aborted take-off.

After repair R.VII flew bombing missions into the summer of 1917 when *Rfa* 501 was transferred to the Western Front. R.VII was then used as a trainer at Vilna into January 1918.

Above: The SSW R.V 5/15.

Above: The SSW R.VI 6/15 after a night landing on a frozen field on December 21, 1916.

Below: The rebuilt SSW R.V 5/15 showing its enlarged wings with supplementary bays added.

Above: Result of a bad landing at night of the SSW R.V on January 31, 1917 that severely damaged the aircraft.

Above: The SSW R.VI as completed with original, short-span wings.

Below: The SSW R.VI as modified with extended-span wings. It was powered by three 200 hp Benz Bz.IV engines.

Above: The SSW R.VI 6/15 on its way to join *Rfa* 501 at Vilna in August 1916.

Right: The SSW R.VI 6/15 in flight.

Below: The SSW R.VI as modified with extended-span wings.

Above: The SSW R.VI 6/15 differed in detail from the R.VII 7/15 below.

Below: The SSW R.VII 7/15 was the first of the series to be powered by 260 hp Mercedes D.IVa engines.

SSW R.VIII & R.IX

Above: The SSW R.VIII was the first of the third-generation SSW R-planes, and was the world's largest airplane when built.

The contract for two R.VIII bombers, 23/16 and 24/16, SSW's final R-plane design to be built, was signed in the summer of 1917. By this time SSW had lost a lot of money on its R-plane program and with it much of its enthusiasm for continued R-plane production. Therefore SSW had changed its focus to fighter aircraft, which were also in great demand and much cheaper and easier to build. However, there was always the hope that its prior R-plane investment could finally bear fruit.

The SSW R.VIII with its 48 meter (157.5 ft.) span was the largest WWI aircraft built; in fact, its span was not exceeded for nearly a decade after the war. The R.VIII was designed to be powered by six coupled engines. Originally 260 hp Mercedes D.IVa engines were planned, but the new 300 hp Basse & Selve BuS.IVa engines were substituted for their greater power. Its required performance was a definite advance over earlier needs; a climb to 4,500 meters in 120 minutes carrying a useful load of 5,250 kg. A modest speed of 130 km/hr was required at 2,500 meters.

A full-scale wood mock-up of the fuselage, wing center-section, gunners' positions, and drive assembly was completed in autumn 1917. However, SSW did not have enough engineering resources to prepare drawings, so *Idflieg* assigned several engineers full-time to the project.

Above: The immense size of the SSW R.VIII is evident in this photo.

The R.VIII was so large that a new assembly hangar had to be built. After it was completed in October 1917, work on 23/16 began immediately. However, once again engine problems delayed the program. The BuS.IVa engines were delayed and there were also problems with the drive system. The result was the R.VIII was still in assembly when the war ended in November 1918.

The engines were arranged in two lines of three each, with the two forward engines coupled together to drive the two-bladed tractor propellers and the four engines at the rear coupled to drive the four-bladed pusher propellers. The R.VIII was designed to cruise on the four forward engines after the bombs had been dropped. All engines could be de-coupled from the gearboxes via clutches. After the cooling problems SSW had experienced with the ill-fated Maybach engines, SSW very carefully designed the oil and water radiators based on flight tests using a Gotha bomber, and these were installed in shrouds based on a Venturi design fitted between the wings.

The R.VIII had a streamlined fairing around the ladder leading to the upper wing machine-gun position, which was designed to mount two machine guns, one on each side. A ventral machine gun was also supplied, and parachutes were stored in the nose and near the rear exit door. Ailerons were fitted to all wings and supplied with Flettner servo-tabs to reduce control forces, the only application of these to an R-plane.

On March 1, 1919 the R.VIII left its hangar under its own power and performed taxi tests. On June 6, 1919, the R.VIII was performing rear engine

tests when the port pusher propeller disintegrated, severely damaging the airframe. On June 26 the German government cancelled repairs on R.23 and completion of R.24, which was about 75% built, and the SSW R.VIII never flew.

On July 24, 1918, three R.VIIIa bombers, R.75–R.77, were ordered. The most important change to the design was addition of a 160 hp Mercedes D.III engine driving a Brown-Boveri supercharger for improved ceiling and speed. Preliminary work started in November but was cancelled with the Armistice.

SSW R.IX

The SSW R.IX was designed as an eight-engine successor to the R.VIII. Assigned military serial number 204/16, the design was changed to a large passenger aircraft after the Armistice. The R.IX was designed for the same 300 hp BuS.IVa engines used in the R.VIII. Other than the conversion of the fuselage for carrying passengers, the R.IX was essentially an enlarged R.VIII, but no construction was undertaken.

Right: Close-up photo of the SSW R.VIII 23/16.

Above: The circular objects attached to the fuselage between the wings were the radiators for the SSW R.VIII.

Below: The enclosed tunnel from the fuselage to the gunner's position on the top wing is shown clearly in this view.

Below: The SSW R.VIII was finished in dark, night-bomber camouflage for its intended role.

Above: The circular objects attached to the fuselage between the wings were the radiators.

Below & Below Right: Close-up views showing R.VIII details and its night-bomber camouflage.

Siemens-Schuckert R-Plane Specifications

	Forssman	R.I	R.II (Orig.)	R.II (Reblt.)	R.III (Orig.)	R.III (Reblt.)
Engines	Mercedes[1]	3 x 150 hp Benz Bz.III	3 x 240 hp Maybach HS	3 x 260 hp Mercedes D.IVa	3 x 240 hp Maybach HS	3 x 200 hp Benz Bz.IV
Span	24 m	28 m	28.22 m	38.0 m	28.22 m	34.33 m
Wing Area	140 m²	138 m²	156 m²	233 m²	156 m²	177 m²
Length	16.5 m	17.5 m	17.7 m	18.5 m	17.7 m	17.7 m
Height	—	5.2 m	4.6 m	4.6 m	4.6 m	4.6 m
Empty Weight	3,250 kg	4,000 kg	5,350 kg	6,150 kg	—	5,400 kg
Loaded Weight	—	5,200 kg	7,150 kg	8,460 kg	—	6,820 kg
Max. Speed:	115 km/h	110 km/h	130 km/h	110 km/h	—	132 km/h
Climb: 2,000m	—	35 min.	—	23 min.	—	35 min.
3,000m	—	—	—	45 min.	—	—
Ceiling:	—	3,700 m	—	3,800 m	—	3,000 m
Duration:	—	4 hours	5¼ hours	4 hours	—	4 hours

Notes: 1. Originally 4x110 hp Mercedes D.I, rebuilt to 2x110 hp Mercedes D.I + 2x220 hp Mercedes D.IV.

Siemens-Schuckert R-Plane Specifications

	R.IV (Orig.)	R.IV (Reblt.)	R.V	R.VI	R.VII	R.VIII
Engines	3 x 240 hp Maybach HS	3 x 200 hp Benz Bz.IV	3 x 200 hp Benz Bz.IV	3 x 200 hp Benz Bz.IV	3 x 260 hp Mercedes D.IVa	6 x 300 hp B&S BuS.IVa
Span	28.22 m	37.6 m	34.33 m	33.36 m	38.44 m	48 m
Wing Area	140 m²	201 m²	177 m²	171 m²	210 m²	440 m²
Length	17.7 m	18.0 m	17.7 m	17.7 m	18.5 m	21.6 m
Height	4.6 m	4.6 m	4.6 m	4.6 m	4.6 m	7.4 m
Empty Weight	—	5,450 kg	5,300 kg	5,250 kg	5,700 kg	10,500 kg
Loaded Weight	—	6,900 kg	6,766 kg	6,800 kg	7,960 kg	15,900 kg
Max. Speed:	—	130 km/h	132 km/h	132 km/h	130 km/h	125 km/h (est.)
Climb: 2,000m	—	36 min.	36 min.	36 min.	27 min.	—
3,000m	—	104 min.	102 min.	—	66 min.	—
Ceiling:	—	3,050 m	3,000 m	2,950 m	3,200 m	4,000 m (est.)
Duration:	—	—	4 hours	5¼ hours	4 hours	—

Right: SSW built 80 Gotha G.IV(SSW) bombers under license in two batches of 40. The Gotha engines were installed as pushers but SSW experimented with the tractor engine configuration on Gotha G.IV(SSW) 211/17 as seen here.

SSW L.I

The SSW L.I was the only bomber built in *Idflieg's* new L-category, a type intermediate between the G and R types. Initially designated the G.III (the G.I and G.II remaining un-built projects), six G.III bombers, (894–899/17) were ordered in October 1917. The type was re-designated L.I in April 1918. The first bomber, L.897/17, was completed in June 1918 and made its initial flight on August 5. It was destroyed in a landing accident a few weeks later. Two more L.I bombers were completed, 898/17 in October 1918 and 899/17 in February 1919.

Seimens-Schuckert L.I Specifications		
Engines:	3 x 240 hp Maybach Mb.IV	
Wing:	Span, Upper	32.0 m
	Wing Area	169 m²
General:	Length	14.65 m
	Empty Weight	4,400 kg
	Loaded Weight	6,400 kg
Maximum Speed:		125 km/h
Endurance:		5.5 hrs.

Above: The SSW L.I was the only aircraft completed to the new L-type category.

Below: The SSW L.I was a large, powerful aircraft that used the tri-motor Caproni-type configuration.

Above: The SSW L.I had a gunner behind the wing in each fuselage boom, and they could fire downward through a Gotha-type tunnel in the tail boom to defend against attacks from directly below.

Above: The SSW L.I 898/17, which was probably the aircraft completed in October 1918. Powered by three 240 hp Maybach Mb.IV engines, the L.I was armed with three flexible machine guns.

The L.I followed the configuration of the Italian Caproni bombers with two fuselage booms and three engines, one in the rear of the short fuselage and two others in the front of the booms. Power was provided by three 240 hp Maybach Mb.IV engines. Each boom carried a gun position and the gunners could fire downward through Gotha-type fuselage tunnels. A third gun position was in the fuselage nose. The Armistice ended development of the L.I before it could be tested in combat

SSW R-Types

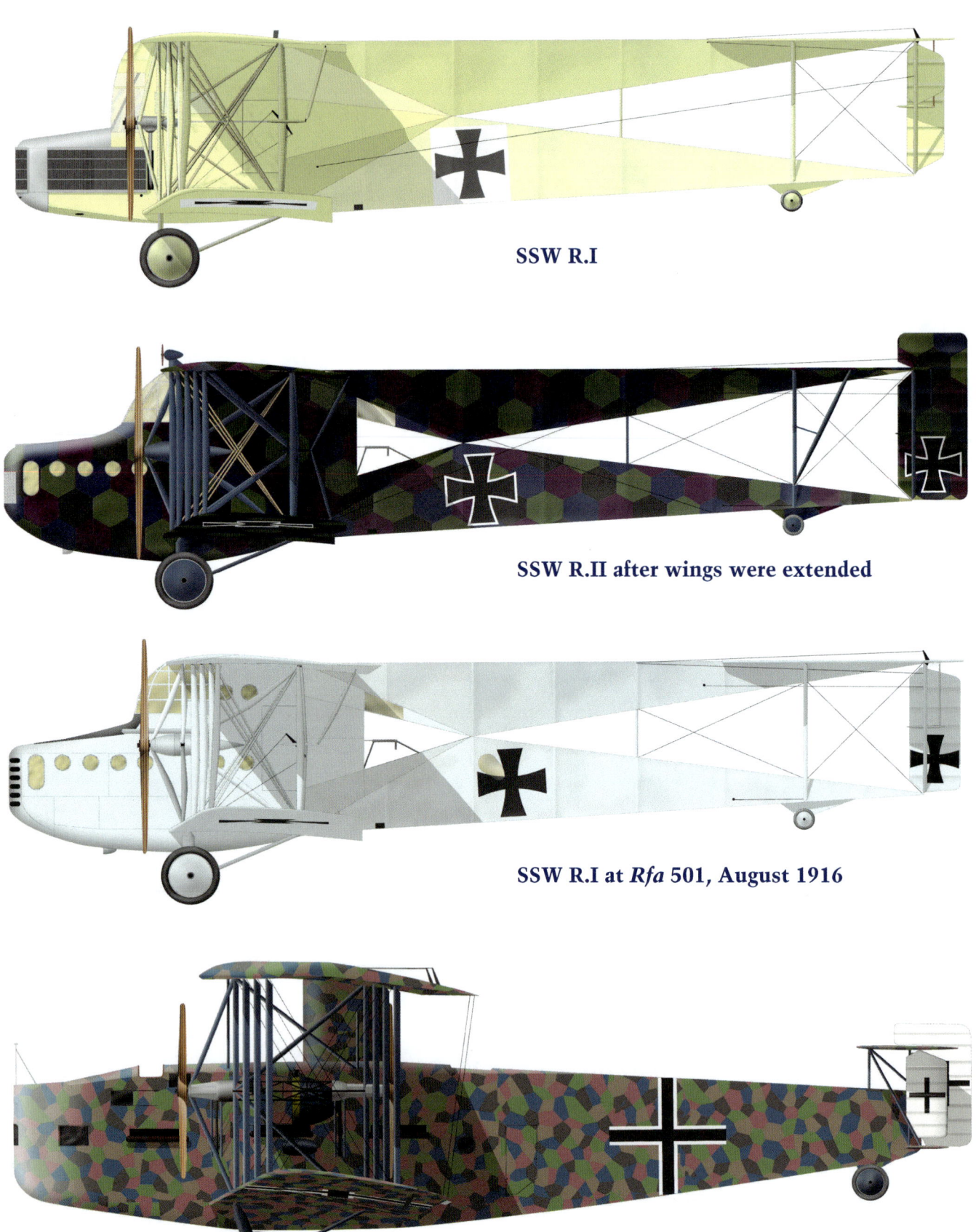

SSW R.I

SSW R.II after wings were extended

SSW R.I at *Rfa* 501, August 1916

SSW R.VIII

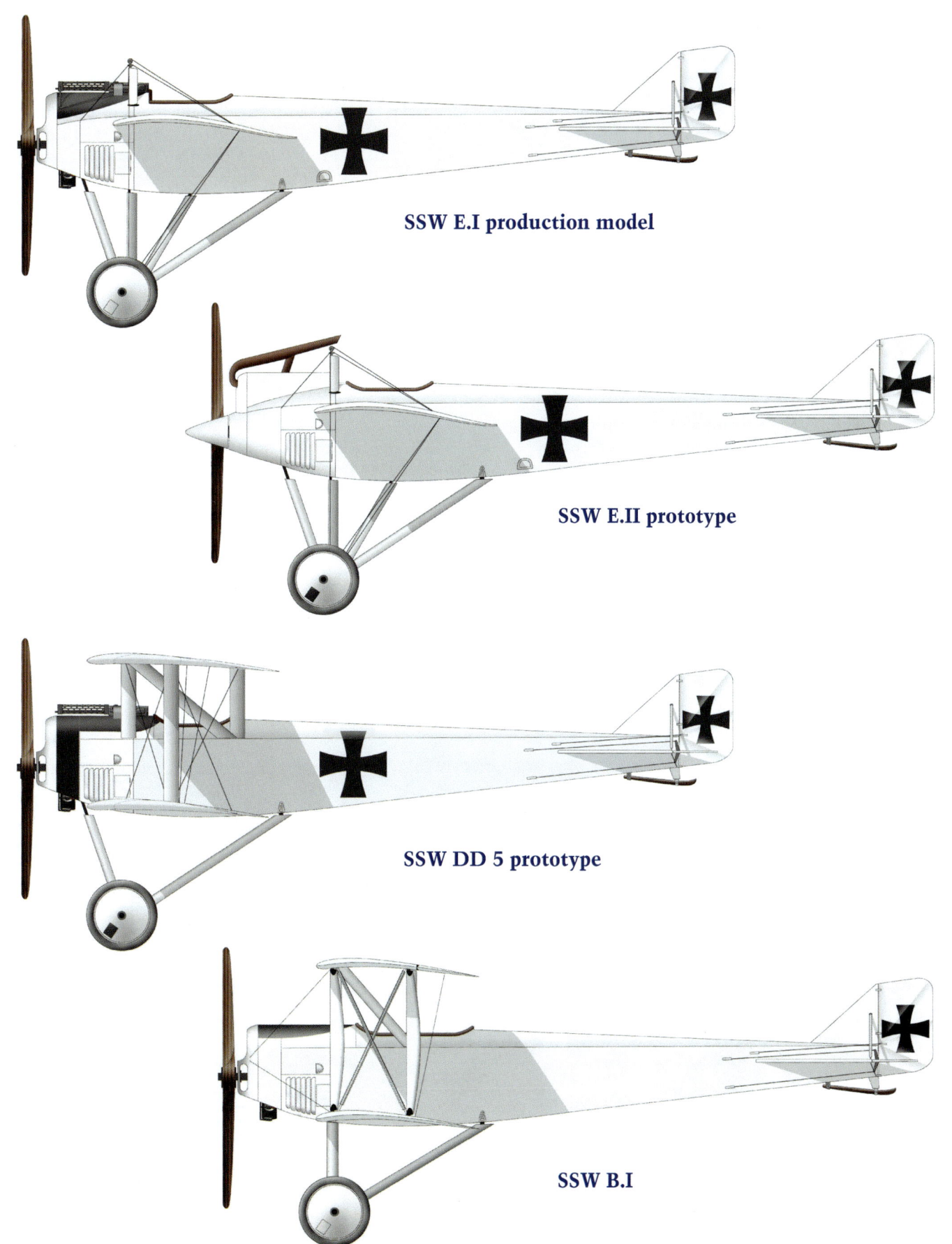

SSW E.I production model

SSW E.II prototype

SSW DD 5 prototype

SSW B.I

SSW Fighters

In parallel with his work on SSW R-planes, Franz Steffen also designed a number of fighters that were built at the SSW Transformer Works in Nürnberg. These started with monoplanes inspired by the Fokker *Eindeckers* that were built in small numbers along with a sesquiplane clearly based on the Nieuport 17 that was designed by Steffen and placed into production as the SSW D.I.

Before Franz Steffen's fatal crash in the SSW E.II he designed, a couple of triplane designs were also tried. This was followed by a series of original biplane fighters, the D.II–D.V, of which the D.III and D.IV were produced in numbers and served successfully in combat. From an operational perspective the D.III and D.IV were by far SSW's most important contribution to the air war. The final SSW fighter, the D.VI, was a parasol monoplane development of the D.IV that offered exceptional performance but arrived too late for production.

SSW E.I

The SSW E.I was completed in October 1915 and finished its tests on March 17, 1916. Of conventional wire-braced wood construction with fabric-covered wings and ply-covered fuselage, it was recommended for production and an order for 20 aircraft, 550–569/15, was placed by *Idflieg*. Production was undertaken by the SSW Transformer Works in Nürnberg. The E.I was fitted with one synchronized machine gun. Eight E.Is were powered by the 90 hp Sh.0 and the remainder by the 110 hp Sh.I. An additional airframe was powered by a 120 hp Argus As.II, becoming the sole SSW E.II. Like the Fokker designs, it used wing-warping instead of conventional ailerons for roll control. Unfortunately, the SSW monoplanes arrived after the D.H.2 and Nieuport 11 had established technical superiority over the early monoplane designs and were too late to make an impact.

Below: The SSW E.I prototype designed by Franz Steffen featured a small spinner that was omitted from production aircraft. Like most *Eindeckers* one synchronized gun was carried and it was powered by a rotary engine, in this case either a 90 hp Sh.0 (eight aircraft) or a 110 hp Siemens-Halske Sh.I counter-rotary engine (a dozen aircraft).

Above & Below: The SSW E.I prototype, identified by the small spinner, photographed at the SSW factory. By the time the SSW *Eindeckers* appeared the basic configuration and wing-warping instead of ailerons, was outdated. (The Peter M. Bowers Collection/The Museum of Flight)

Above: The fuselage of the SSW E.I in front of the pilot was painted a dark color, probably to reduce glare.

Below: SSW E.I fighter airframe undergoing static load testing at the SSW factory. The load had to be distributed along the wing to simulate flight loads. The sandbag method was used widely throughout the aviation industry for many years.

Above: The SSW E.I production aircraft dispensed with the small spinner fitted to the prototype.

Below: SSW E.I fighters in production at the modern SSW factory. The SSW design was cleaner than the Fokker and Pfalz *Eindeckers* but was simply too late for important operational use at the front and most were delivered to single-seat flying schools.

Above: Another view of an SSW E.I production aircraft. The support structure for the front bearing needed by the Sh.I counter-rotary engine is clearly shown. (The Peter M. Bowers Collection/The Museum of Flight)

Below: This close-up view of an SSW E.I fighter shows the gun installation in greater detail together with a closer view of the dark-painted anti-glare panel in front of the pilot.

SSW E.I Specifications		
Engine:	110 hp Siemens-Halske Sh.I	
Wing:	Span	10.00 m
	Area	16.0 m²
General:	Length	7.10 m
	Height	2.80 m
	Empty Weight	473 kg
	Loaded Weight	673 kg
Maximum Speed:		140 km/h

Above: An SSW E.I production aircraft in flight. Although using the same configuration as the Fokker and Pfalz *Eindeckers*, the SSW E.I was much different in detail, and in particular the tail surfaces were distinctly different. A review of the SSW *Eindecker* delivery data on page 56 shows that most of these airplanes were delivered directly to flying schools.

SSW E.II

The SSW E.II was essentially an E.I airframe powered by an in-line 120 hp Argus As.II engine; like the E.I it carried a single machine gun. The one prototype built first flew in early 1916. Its designer, Franz Steffen, suffered a fatal crash in the E.II while demonstrating it at Döberitz on June 26, 1916.

Above: The sole SSW E.II prototype is shown in side view. The large exhaust for its in-line Argus engine is prominent. The SSW E.II was one of two *Eindeckers* to feature an in-line engine, the other being the Pfalz E.V.

Above: The SSW E.II prototype under construction. This view clearly shows the well-streamlined installation of the in-line 120 hp Argus As.II engine. (The Peter M. Bowers Collection/The Museum of Flight)

Below: Rear view of the SSW E.II fighter prototype at the SSW factory. For an *Eindecker* the E.II was quite streamlined.

Above: The SSW E.II prototype designed by Franz Steffen was powered by a 120 hp Argus As.II engine. Like most *Eindeckers*, one synchronized gun was carried. (The Peter M. Bowers Collection/The Museum of Flight)

Below: The fatal crash of Franz Steffen in the SSW E.II prototype he designed. Steffen was demonstrating the E.II at Döberitz on June 26, 1916 when he crashed.

SSW E.III & E.IV

Ordered in April 1916, the SSW E.III was an E.I airframe with a 100 hp Oberursel; again one synchronized gun was carried. Six aircraft, 620–625/16, were built. A derivative of the E.III was proposed with a circular fuselage; known as the E.IV, it remained a project.

Siemens-Schuckert Monoplane Fighter Specifications

	E.I	E.II	E.III	D.VI
Engine	110 hp Sh.I	120 hp As.II	100 hp Oberursel U.I	205 hp Sh.IIIa
Armament	One MG	One MG	One MG	Two MG *
Span	10.00 m	—	7.10 m	9.37 m
Length	7.10 m	—	7.10 m	6.50 m
Height	2.80 m	—	2.80 m	2.72 m
Wing Area	16.0 m²	—	16.0 m²	12.46 m²
Empty Weight	473 kg	—	478 kg	540 kg
Loaded Weight	673 kg	—	678 kg	710 kg
Maximum Speed:	140 km/h	—	—	220 km/h (@ 2,000m)
Climb: 6,000m	NA	—	—	16 minutes
7,000m	NA	—	—	22 minutes
Ceiling:	—	—	—	8,000m

Notes:1. E.II dimensions similar to E.I except length.
2. SSW D.VI armament is that projected for production; the two prototypes were unarmed.

SSW E-Type Deliveries

Type	Military #	Date	Delivered To
E.I	550/15	August 1916	R-*Abteilung Rfa* 501
E.I	551/15	August 1916	R-*Abteilung Rfa* 500
E.I	552/15	August 1916	Sent to Adlershof as static test airframe
E.I	553/15	September 1916	*Armeeflugpark* 8
E.I	554/15	September 1916	*Armeeflugpark* 8
E.I	555/15	September 1916	*Armeeflugpark* 8
E.I	556/15	Oct. 9, 1916	*Armeeflugpark* 10
E.I	557/15	Dec. 29, 1916	*Kampfeinsitzerschule* I, *Fea* 7, Köln
E.I	558/15	Dec. 14, 1916	*Kampfeinsitzerschule* I, *Fea* 7, Köln
E.I	559/15	Jan. 10, 1917	*Kampfeinsitzerschule* I, *Fea* Köln
E.I	560/15	Jan. 20, 1917	*Kampfeinsitzerschule* I, Paderborn
E.I	563/15	Jan. 13, 1917	*Kampfeinsitzerschule* I, *Fea* Köln
E.I	564/15	Jan. 20, 1917	*Kampfeinsitzerschule* I, Paderborn
E.I	561, 562, 565–569/15	February 1917	These aircraft were sent to Adlershof for cannibalization
E.II	—	Early 1916	Destroyed in crash June 26, 1916
E.III	620/16	Sept. 9, 1916	*Armeeflugpark* 8
E.III	621/16	Dec. 14, 1916	*Kampfeinsitzerschule* I, *Fea* 7, Köln
E.III	622/16	Dec. 14, 1916	*Kampfeinsitzerschule* I, *Fea* 7, Köln
E.III	623/16	Dec. 14, 1916	*Kampfeinsitzerschule* I, *Fea* 7, Köln
E.III	624/16	Dec. 14, 1916	*Kampfeinsitzerschule* I, *Fea* 7, Köln
E.III	625/16	Dec. 29, 1916	*Kampfeinsitzerschule* I, *Fea* 7, Köln

Note: Data from SSW archives via Peter Grosz and Dick Bennett

SSW DD 5

Designed by Franz Steffen, the SSW DD 5 was basically an E.I airframe with a biplane wing cellule, and was SSW's first biplane fighter. Power remained a 110 hp Siemens-Halske Sh.I counter-rotary engine, and a single synchronized machine gun was fitted. Tubular steel was used for the wing spars and I-struts. Tested in August 1916, it was rejected because of poor field of view for the pilot and poor handling qualities. Development then focused on the SSW D.I.

Unusually for a WWI aircraft, the DD 5's wing planform tapered from the fuselage to the wing tips; most WWI featured wings of rectangular planform as seen in the SSW E-types. Tapered wings are more efficient because they generate less induced drag (drag induced by creating lift, which generates vortices at the wing tips) and also reduce weight,

which is likely why the designer used taper.

However, tapered wings aggravate the tendency of the wing to stall first at the tips. Since the ailerons are normally fitted at the tips of the wings for greatest effectiveness, the tendency of the wing to stall first at the tips is very dangerous because it robs the pilot of roll control in a critical situation, increasing the likelihood of the aircraft entering an unintended spin. Most designers of the time 'washed out' the angle of incidence at the wing tips to prevent the wing from stalling at the tip first, and more sophisticated means were developed later to prevent tip stall. However, a look at the DD 5 shows no evidence of any aerodynamic measures designed into the wing to alleviate the tip stall problem, undoubtedly resulting in poor handling near stalling speed.

Above: The SSW DD 5 prototype designed by Franz Steffen was SSW's first biplane fighter design. The wing spars and I-struts were made of steel tubes, and the aircraft has a robust appearance despite the narrow I-struts. Other than the biplane wing cellule, the rest of the aircraft was derived from the E.I monoplane fighter, including its 110 hp Siemens-Halske Sh.I counter-rotary engine and single synchronized gun. The DD 5 was rejected due to poor handling qualities, probably related to its tapered wing planform that aggravated the tendency of the wing tips to stall first, which seriously reduced the pilot's ability to control aircraft roll during a critical flight regime.

Above: The SSW DD 5 prototype in the foreground and E.I in the background emphasize their similarity in design.

Below: The SSW DD 5 was a clean design but the wide cabane struts near the pilot must have restricted his field of view.

SSW D.I

During the spring and summer of 1916 the Nieuport 11 and D.H.2 biplanes wrested air superiority from the Fokker and Pfalz monoplanes due to their superior performance and maneuverability. Since the new Halberstadt, Fokker, and Albatros biplane fighters had barely started to arrive at the Front and had not yet proved themselves, *Idflieg* decided that one way to quickly redress the balance would be to produce improved copies of the Nieuport fighters, which clearly had more performance and development potential than the D.H.2 pusher. *Idflieg* therefore provided captured Nieuport fighters to the Albatros, Euler, Fokker, Pfalz, and Siemens-Schuckert companies and encouraged them to create new fighters based on Nieuport technology.

Albatros adapted the Nieuport sesquiplane wing design to their new D.II fighter to develop their Albatros D.III and later D.V and D.Va fighters. Pfalz adopted the wing planform for their D.III fighter but retained the stronger two-spar lower wing, avoiding the structural problems experienced by Nieuport and Albatros. Euler produced a near copy of the Nieuport, while Fokker pursued his own designs. Siemens-Schuckert, already involved in building giant bombers, chose to reduce development time by producing a near copy of the Nieuport.

Other than its use of a 9-cylinder, 110 hp Siemens-Halske Sh.I rotary engine, the SSW response, designed by Franz Steffen shortly before his fatal crash in the E.II and designated the SSW D.I, was essentially a copy of the Nieuport 17. The engine cowling and propeller differed from the Nieuport due to the different engine. The slow-turning Sh.I enabled a larger, more efficient propeller to be used, which in turn required a longer undercarriage for clearance. The Sh.I also required front bearing supports in the cowling. The center section appeared a little more compact with slightly reduced gap between the wings, giving the SSW D.I a more aggressive appearance than the Nieuport. Wing area was slightly less than the Nieuport original, and a single synchronized Spandau machine gun was mounted just to the right of centerline ahead of the pilot. After the first few production machines were built, a spinner was added to improve streamlining.

The prototype SSW D.I, 3503/16, was first test flown between 4–10 October 1916 by Bruno Steffen; during tests it climbed to 5,000 meters in 45 minutes, a reasonable performance for the time. The *Typenprüfung* (type test) was flown on October 26, 1916 and the SSW D.I was accepted pending successful static load testing. The second D.I, 3504/16, was tested to destruction during 25–27

Below: A mid-production SSW D.I serving with *Jasta* 7 sports a spinner. Together with the reduced gap, the spinner gives the D.I a more aggressive look than the Nieuport 17 on which it was based.

SSW D.I Production Orders				
Order Date	**Siemens #**	**Quantity**	**Military #**	**Notes**
25 Nov. 1916	9733	1	D3503/16	
25 Nov. 1916	9762	4	D3504 – 3507/16	
25 Nov. 1916	9772	6	D3508 – 3513/16	
25 Nov. 1916	9798	122	D3514 – 3635/16	
25 Nov. 1916	9777	16	D3552 – 3767/16	
25 Nov. 1916	9866	1	D3768/16	D.Ia
21 March 1917	9899	1	D1230/17	D.Ib
21 March 1917	9916	1	D1231/17	D.Ib
21 March 1917	9922	98	D1232 – 1329/17	D.Ib, not built
A total of 250 D.I aircraft were ordered, the initial order (206615 Kr.16) for 150 being placed on 25 November 1916 and the second order for 100 aircraft (371724 Kr.17) on 21 March 1917.				

January 1917.

A total of 250 D.I aircraft were ordered, the initial order (206615 Kr.16) for 150 being placed on November 25, 1916 and the second order for 100 aircraft (371724 Kr.17) on March 21, 1917. The order dates, Siemens order numbers, quantities, and associated military serial numbers are given in the table above.

Below: The prototype SSW D.I, 3503/16, in the factory with an R-plane under construction in the background.

Above: The prototype SSW D.I, 3503/16, wears small *Eiserneskreuz* markings on the upper wings reminiscent of the E.I marking style. At right appears to be a Nieuport 16 fuselage and engine cowl.

However, only 95 of the original order were completed (22 at Berlin and 73 at Nürnberg) as production was seriously delayed by slow engine deliveries. In the meantime the Albatros D.II and D.III, with their more reliable and powerful engines and twin Spandau guns, had demonstrated their superiority. Therefore SSW D.I production was terminated by *Idflieg* in July 1917. The Nürnberg plant shipped an additional 55 partially completed airframes to Adlershof for instructional use in aviation mechanics' schools.

Photographs show the various D.I production batches had some distinguishing features:

Prototype: No louvers, no spinner, straight tailskid, light plain finish, no camouflage. Small crosses on white fields on top wing surface, underneath lower wing surface, and on fuselage. Cross on rudder also on white field. Engine air intake on starboard side

Early Production: Four louvers, no spinner, fabric covered center section of top wing, curved tailskid, three-color camouflage, crosses on white fields, crosses under upper wings, engine air intake on port side, rigging diagram on port side below cockpit.

Mid-Production: Three louvers, spinner, plywood covered center section of top wing, curved tailskid, two-color camouflage, crosses with borders, no crosses under upper wings, engine air intake on starboard side.

Late Production: Three louvers, spinner more sharply pointed than earlier spinner, plywood covered center section of top wing, outer-braced tailskid, two-color camouflage, crosses with borders, no crosses under top wings, engine air intake on starboard side.

Above: The prototype SSW D.I with four-bladed propeller and Sh.I engine. The machine gun is offset slightly to the right.

The SSW D.I was used only in small numbers and no *Jasta* was ever fully equipped with the type. Units which received the SSW D.I included *Jastas* 1, 2, 3, 4, 5, 7, 9, 11, and 14; two went to *Armee Flugpark Süd* in Galicia. The D.I had excellent flying qualities (*Lt.* Kurt Student praised it as having the best flying qualities of any fighter at the front at that time) but modest performance, so many were delivered to fighter schools.

SSW D.I Development

A number of modifications were tried to improve D.I performance, including enlarged wing area and up-rated engines. D.I 3576/16 was tested with a Siemens-designed supercharger, making several successful factory flights in August–September 1918 before going to Adlershof for additional testing. Three aircraft with more wing area were built at the Berlin factory. First was SSW D.Ia (3768/16) with two guns and 15.7 m² wing area; delivered in July 1917, it had an empty weight of 430 kg and

reached 4,000 meters in 25 minutes. There were two variants of the D.Ib, both with two guns, one-piece upper wings, and a length of 6 meters. The first D.Ib, 1230/17, had a span of 8.7 m, wing area of 16.2 m², empty weight of 410 kg, and reached 5,000 meters in 29.8 minutes. The second D.Ib, 1231/17, had a larger wing spanning 11.1 m with 19.2 m² wing area, empty weight of 430 kg, and an over-compressed 140 hp Sh.Ia engine. It demonstrated excellent climb, reaching 5,000 meters in 20.5 minutes, less than half the time required by the D.I.

Despite its excellent climb, the D.Ib was not fast enough to see production in late 1917. The basic Nieuport sesquiplane configuration had reached the limit of its development potential, and subsequent SSW fighters featured conventional two-spar wings and the more powerful 11-cylinder Sh.III engine.

The proposed D.Ic, a parasol monoplane with 8.2 meters span, 560 kg empty weight, and 160 hp Sh.III engine remained a project, but inspired the later SSW D.VI parasol monoplane fighter.

Above & Below: The prototype SSW D.I, 3503/16, at Döberitz with designer and test pilot Bruno Steffen, the D.I designer's brother. The Sh.I counter-rotary was powerful and its propeller was slow-turning, resulting in a large propeller diameter and tall undercarriage when a two-blade propeller was fitted. The Sh.I required bearing supports at the front of the cowling. There is no carburetor intake scope on the port side.

Above: The prototype SSW D.I, 3503/16, shows its carburetor intake scoop on the starboard side and no side vents on the engine cowling.

Below: The first six SSW D.I on the SSW factory airfield at Nürnberg. Only the second aircraft in line has the three-color camouflage used on some of the aircraft.

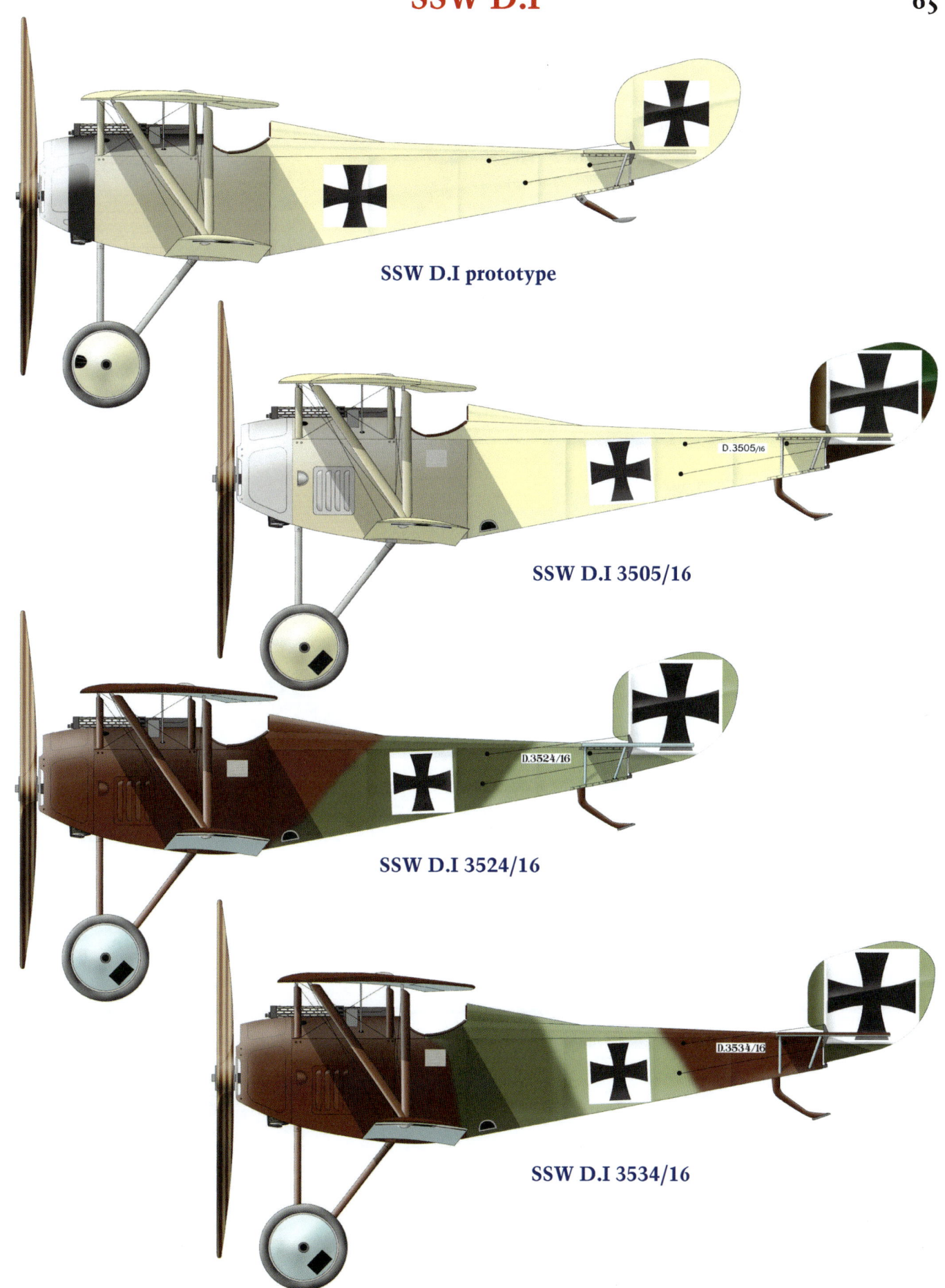

SSW D.I prototype

SSW D.I 3505/16

SSW D.I 3524/16

SSW D.I 3534/16

SSW D.I

SSW D.I 3506/16 of *Jasta 7*

SSW D.I 3760/16 of *Jasta 14*

SSW D.I 3761/16 of *Jasta 5*

SSW D.I 3767/16

Right & Below: SSW D.I 3511/16, an early production machine, in slightly different views. The white rectangle under the cockpit carries the rigging instructions. This aircraft was sent to *Jasta* 11 via *Armeeflugpark* 6 on May 7, 1917, for *Lt.* Karl Emil Schäfer. Schäfer had 25 victories at the time and had been awarded the *Pour le Mérite.* Schäfer increased his score to 30 before being killed in action on June 5.

Below: SSW D.I 3513/16, an early production machine, wears the three-color camouflage scheme and carries large *Eiserneskreuz* markings on white fields underneath the upper as well as the lower wings. The large propeller needed to absorb the power from the slow-turning 110 hp Sh.I engine is prominent.

Above: SSW D.I 3506/16 was shipped to *Jasta* 7 on Feb. 16, 1917.

Below: Another view of SSW D.I 3506/16 at *Jasta* 7 in February 1917. The slow-turning Sh.I counter-rotary engine required a large diameter propeller for best efficiency, which in turn required a taller landing gear than the Nieuport 17 on which its design was based. The propeller spinner was also different in shape than the fixed, hemispherical cone of penetration fitted to a few Nieuports.

Above: This SSW D.I appears to be in service at the front but could be at a flying school.

Below: This SSW D.I was assigned to *Jasta* 5. From the delivery data, this is probably 3761/16 shipped on April 25, 1916. A cloth covering protects the propeller and no spinner is fitted, probably because of the protective cloth covering.

Above: SSW D.I 3761/16, with spinner now fitted, was assigned to *Jasta 5. Lt.* Kurt Schneider is in the cockpit.

Below: Another view of SSW D.I 3761/16 at *Jasta* 5 with *Hptm.* Hans von Hühnerbein in the cockpit.

Right & Below: The smoldering remains of SSW D.I 3761/16 after being fatally crashed by *Hptm.* Hans von Hühnerbein at *Jasta* 5 at Boistrancourt airfield on May 7, 1917, during a test flight. Hühnerbein was *Staffelführer* of *Jasta* 5 and had scored his only victory on April 7, 1917, ironically over a Nieuport 17.

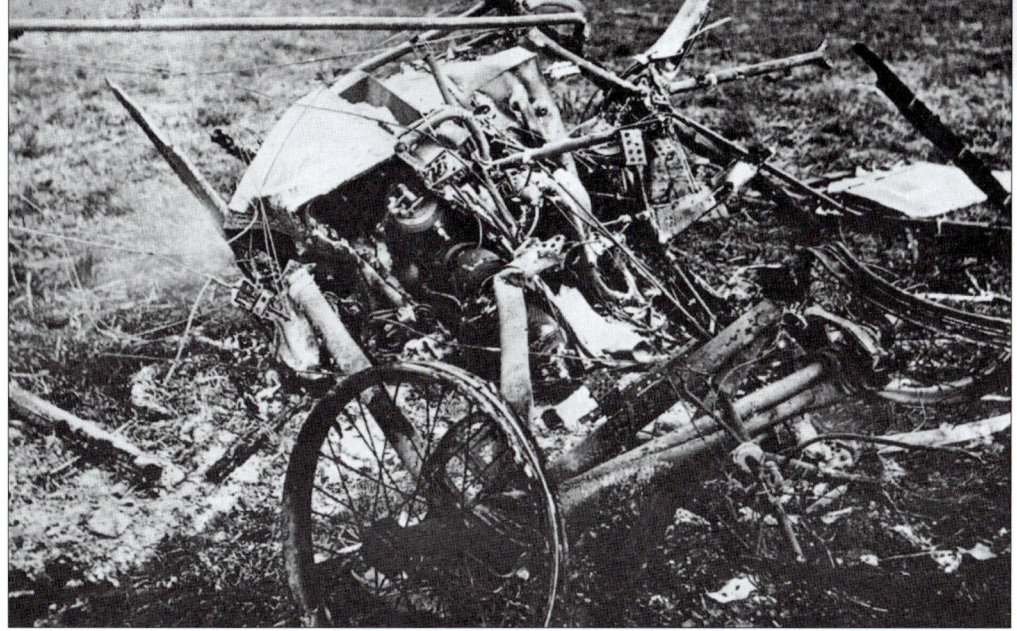

Right: This SSW D.I crashed at the SSW factory, fortunately with less dramatic results than the crash of D.I 3761/16.

Above & Below: This SSW D.I assigned to *Jasta* 14 apparently suffered a bad landing. From the delivery data this could be 3505/16 or possibly 3760/16.

Above & Below: SSW D.I 3767/16 was one of the last D.I fighters built and was sent to the *Flugzeugmeisterie* at Adlershof along with five other D.I fighters and D.Ia 3768/16. The upper wing was close to the fuselage to provide the pilot with a good field of view over the wing. The small gap, the headrest, and the large propeller with sizable spinner all gave the SSW D.I a more aggressive look than the Nieuport on which it was based. The tail skid differed from that on early machines.

Above & Below: These photos of a number of SSW D.I fighters were probably taken at the factory; no *Jasta* ever had this many SSW D.I aircraft assigned to it.

SSW L.I prototype

Above: SSW D.I fighters probably at the factory and photographed at the same time as the lineups on the facing page.

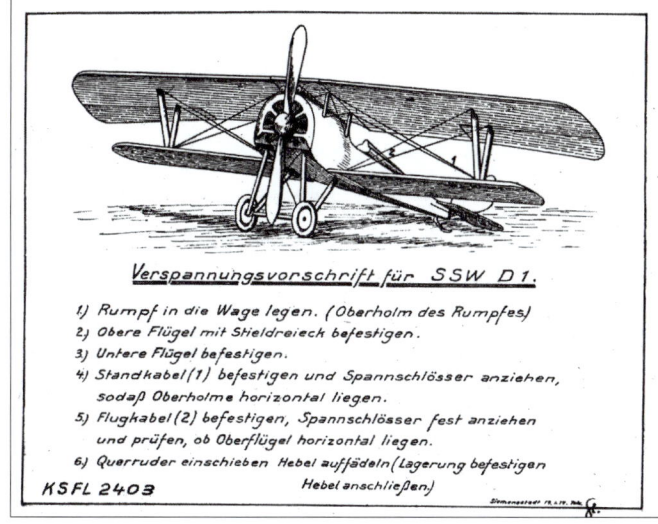

Above: SSW D.I rigging diagram.

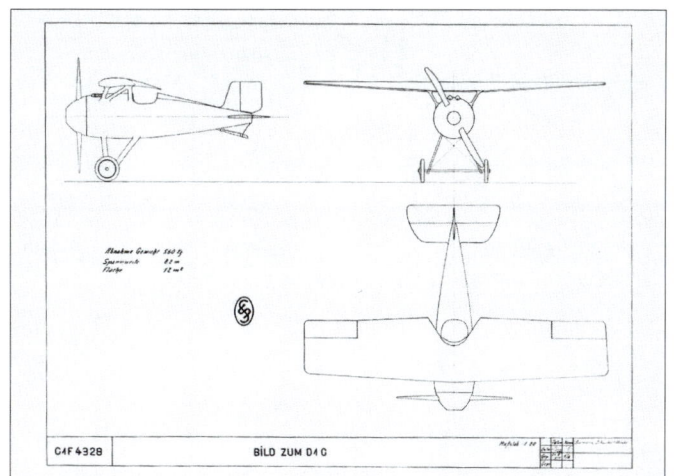

Above: SSW D.Ic proposal. Although disapproved when submitted during the height of *Idflieg's* triplane craze, it foreshadowed the later SSW D.VI parasol monoplane.

**Initial SSW D.II Design
Based on the D.I; This
Aircraft Was Not Built**

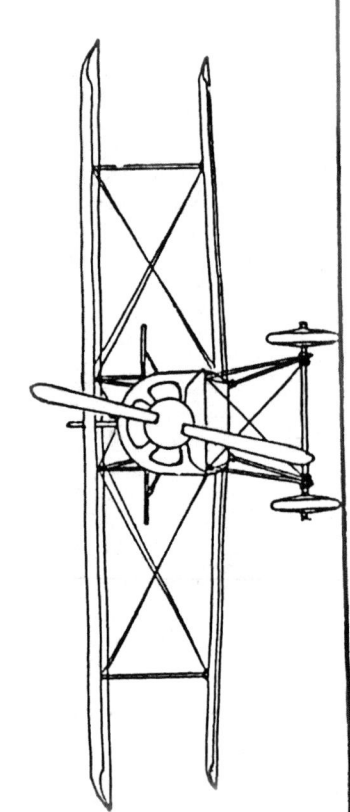

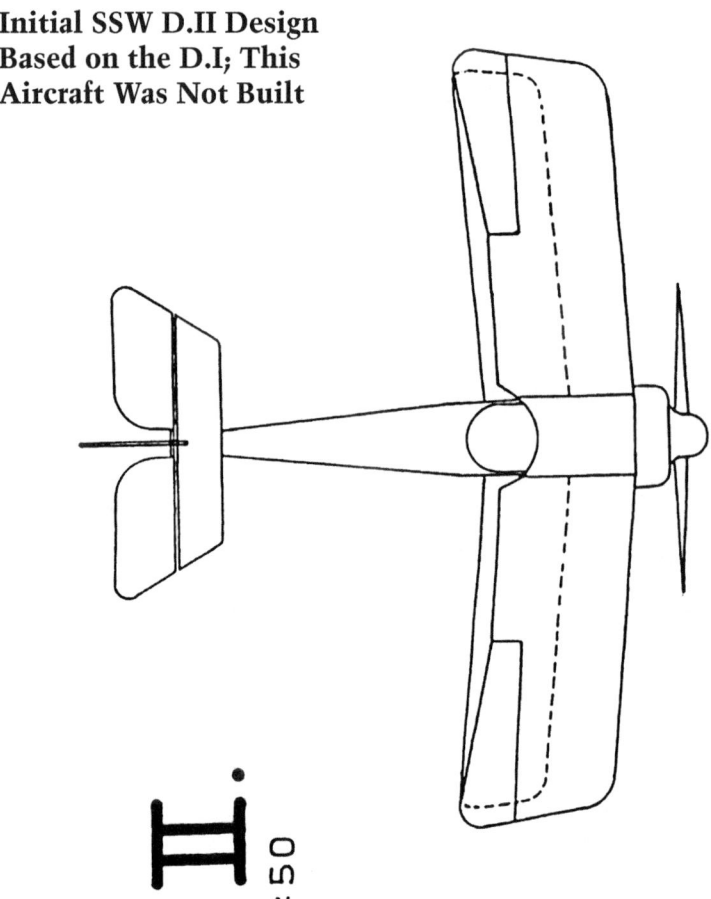

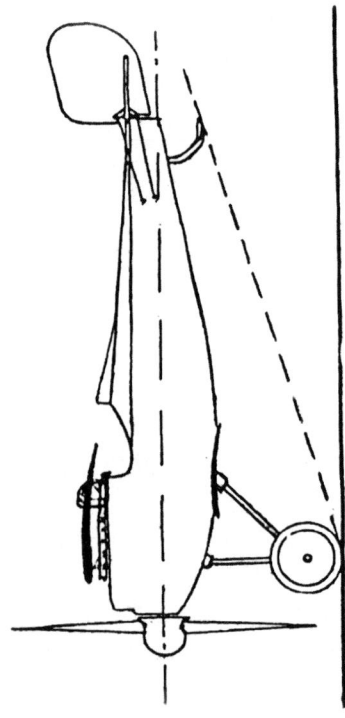

D II.
1:50

MOTORLEISTUNG · · · · N_o = 160 PS

FLUGGEWICHT · · · · · G = 750 Kg

TRAGFLÄCHE · · · · · · · F = 18,7 m^2

SPANNWEITE OBEN · · b_o = 7,80 m

       UNTEN · b_u = 7,39 "

TIEFE OBEN · · · · · · · · t_o = 1,60 "

  " UNTEN · · · · · · t_u = 0,90 m

PROFIL: OBEN 450. UNTEN 138.

**Initial SSW D.IIb Design
Based on the D.I; This
Aircraft Was Not Built**

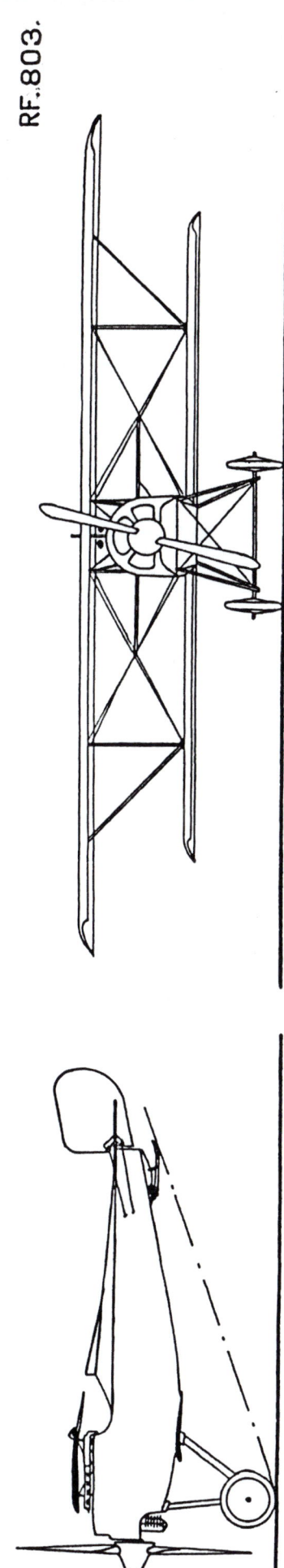

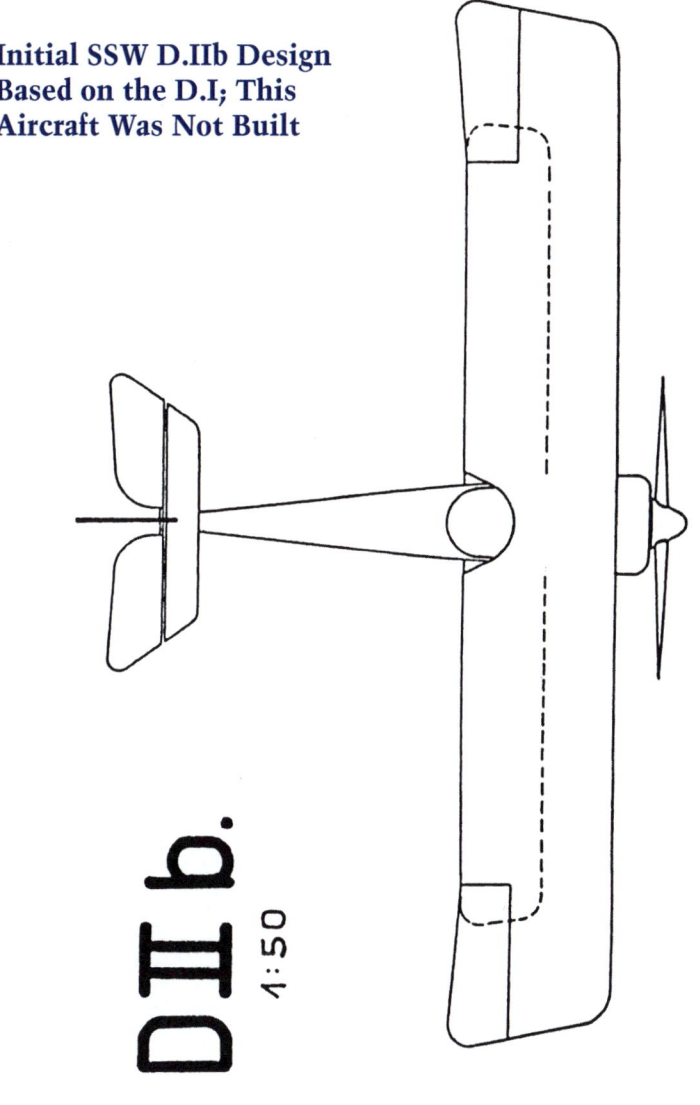

D II b.
1:50

MOTORLEISTUNG . . . . . $N_o = \frac{160}{190}$ PS

FLUGGEWICHT . . . . $G = 700$ Kg

TRAGFLÄCHE . . . . $F = 19,4\ m^2$

SPANNWEITE OBEN . . $b_o = 9,64\ m$

v   UNTEN . . $b_u = 7,00$ "

TIEFE OBEN . . . . $t_o = 1,60$ "

"   UNTEN . . . . $t_u = 0,80\ m$

PROFIL: OBEN 455, UNTEN 181.

Known SSW D.I Deliveries			
Type	**Military #**	**Date**	**Delivered To**
D.I	3503/16	Oct. 26, 1916	Sent to Adlershof as type test aircraft
D.I	3504/16	January 1917	Sent to Adlershof as static test airframe
D.I	3505/16	May 14, 1917	Sent to *Jasta* 14
D.I	3506/16	Feb. 16, 1917	Sent to *Jasta* 7, *Armeeflugpark* 5
D.I	3507/16	Feb. 17, 1917	Sent to *Jasta* 11
D.I	3511/16	May 7, 1917	Sent to *Jasta* 11, *Armeeflugpark* 6, for *Lt.* Schäfer
D.I	3517/16	?	Sent to *Jasta* 9, *Park* Rethel
D.I	3524/16	July 4, 1917	*Militärfliegerschule 5*, Gersthofen
D.I	3531/16	July 4, 1917	*Militärfliegerschule 5*, Gersthofen
D.I	3534/16	July 4, 1917	*Militärfliegerschule 5*, Gersthofen
D.I	3553/16	July 4, 1917	*Militärfliegerschule 5*, Gersthofen
D.I	3554/16	July 4, 1917	*Militärfliegerschule 5*, Gersthofen
D.I	3752/16	March 20, 1917	Sent to *Jasta* 9, *Armeeflugpark* 3. In June 1917 as SSW redeemed the interrupted schedule.
D.I	3753/16	March 27, 1917	Sent to *Jasta* Boelcke (aka *Jasta* 2)
D.I	3754/16	April 12, 1917	Sent to *Jasta* Boelcke (aka *Jasta* 2)
D.I	3755/16	1917	Sent to Adlershof for testing & evaluation
D.I	3756/16	May 18, 1917	*Armeeflugpark Süd*
D.I	3757/16	May 8, 1917	*Armeeflugpark Süd*
D.I	3758/16	April 25, 1917	Sent to *Jasta* 1
D.I	3759/16	April 28, 1917	Sent to *Jasta* 3
D.I	3760/16	April 24, 1917	Sent to *Jasta* 4 (or 14)
D.I	3761/16	April 25, 1917	Sent to *Jasta* 5, *Armeeflugpark* 2
D.I	3762/16	Sept. 19, 1917	Sent to *Flugzeugmeisterie* at Adlershof
D.I	3763/16	Sept. 19, 1917	Sent to *Flugzeugmeisterie* at Adlershof
D.I	3764/16	Sept. 19, 1917	Sent to *Flugzeugmeisterie* at Adlershof
D.I	3765/16	Aug. 17, 1917	Sent to *Flugzeugmeisterie* at Adlershof
D.I	3766/16	1917	Sent to *Flugzeugmeisterie* at Adlershof
D.I	3767/16	1917	Sent to *Flugzeugmeisterie* at Adlershof
D.Ia	3768/16	July 1917	Sent to *Flugzeugmeisterie* at Adlershof
D.Ib (16)	1230/17	Sept. 22, 1917	Sent to Adlershof for testing & evaluation
D.Ib (19)	1231/17	Nov. 12, 1917	Sent to Adlershof for testing & evaluation

Notes:
1. Data from SSW archives via Peter Grosz and Dick Bennett.
2. SSW D.IIa was serial 3500/16, SSW D.II was 3501/16, and SSW D.IIb was 3502/16. The SSW chart states in error that these were all xxx/**17** serial numbers.
3. The SSW Dr.I was assigned serials 3053/17, 3054/17, and 3055/17. Only the first serial was used for the sole Dr.I; the last two serials were re-assigned to the SSW D.VI parasol monoplane prototypes.

Above: An early-production SSW D.I with no spinner and four cooling louvers on the fuselage side.

SSW Dr.I & Dr.II

As might be expected during the German Triplane Craze of 1917, SSW also built a triplane fighter. The fuselage of SSW D.I 3752/16 was used together with its 110 hp Siemens-Halske Sh.I counter-rotary engine and it received the new number 3053/17. Initially it had 8.6 m span and 5.3 m length, with an empty weight of only 425 kg and maximum weight of 632 kg. After a crash during flight testing it was rebuilt with larger wings (details unknown) and weighed more; empty weight was now 500 kg and

maximum weight was 695 kg. It reached 4,700 m in 20.6 minutes. Unfortunately, no photographs of this aircraft have been found despite the fact that it was completed in July 1917 and flew for some time.

A second triplane design, the Dr.II, with more powerful 160 hp Siemens-Halske Sh.III and using a D.IIb fuselage, was scrapped during construction. Serials were assigned for two aircraft, 3054–3055/17; these were later applied to the SSW D.VI prototypes.

SSW DDr.I & DDr.II

In June 1917, at the height of *Idflieg's* 'Triplane Craze', SSW presented the DDr.I design to *Idflieg*, which quickly approved it. The "Flying Egg" was powered by two tandem 125 hp Siemens-Halske Sh.Ia engines and fitted with two synchronized machine guns, then the standard German fighter armament.

Insufficiently stable, the DDr.I crashed on its first flight in November 1917. This eventually resulted

Seimens-Schuckert DDr.I Specifications		
Engines:	2x125 hp Siemens-Halske Sh.Ia	
Wing:	Span, Upper	7.20 m
	Wing Area	18.1 m²
General:	Length	5.3 m
	Empty Weight	510 kg
	Loaded Weight	695 kg
Climb: Dr.IIa:	4,700m	20.6 min

Above: The SSW DDr.I used two high-compression 125 hp Sh.Ia engines, one mounted in the front of the nacelle and the other mounted in the rear of the nacelle. Two synchronized guns were fitted. The eccentric design crashed on its first test flight before performance data could be recorded.

in cancellation of the DDr.I and a proposed, more powerful derivative, the DDr.II powered by two 160 hp Siemens-Halske Sh.III engines.

Centerline thrust can provide improved performance compared to conventional twin-engine designs while eliminating asymmetric thrust problems in case of engine failure. The later Dornier Do-335 is a good example of the potential of a properly designed centerline thrust aircraft; it was significantly faster than similar aircraft of more conventional layout. However, the 'lattice-tail' design of the DDr.I had too much drag, which

Below: The SSW DDr.I after its first test flight. Despite this setback, it was rebuilt, the wings being enlarged, and testing continued for a time.

Above: The two 125 hp Sh.Ia engines powering the SSW DDr.I gave it a good power to weight ratio, but the design of the strut-braced tail created too much drag. The triplane configuration, while providing good lift, also created too much drag.

was surely aggravated by its triplane wing cellule. Even if the DDr.I had acceptable flying qualities it would likely have been too slow, although it might have demonstrated a good rate of climb. Use of two engines in a fighter would also have aggravated Germany's chronic engine shortage.

SSW DDr.I Triplane
Fighter Prototype

SSW D.II to D.V

Above: SSW D.III 1626/18 of *Kest* 4b flown by *Vzfw*. Reimann. Powered by the 205 hp Siemens-Halske Sh.III, the D.III was very maneuverable and had an outstanding rate of climb. The need for more speed led to the similar and faster D.IV.

The SSW D.II through D.V are discussed together because all were variations on the same basic design. Although there were many detail differences, the key distinctions between types were the wings.

Expecting the eleven-cylinder 160 hp Sh.III engine Siemens-Halske was developing, in November 1916 *Idflieg* ordered three D.II prototypes (D.IIa 3500/16, D.II 3501/16, and D.IIb 3502/16) for flight trials.

Two D.II fighters, completed in January 1917, gathered dust at the factory until early June when a Sh.III flight-test engine was at last delivered. SSW quickly installed the new engine and flight trials began on June 7. After a period of laborious experimentation with both two- and four-bladed propellers to maximize performance, *Lt*. Hans Müller, a 12-victory ace and company test pilot, performed an astonishing climb of 7,000 meters in 35.5 minutes on August 5, 1917. Müller flew the D.IIb prototype powered by an experimental, over-compressed Sh.III engine. Claiming an unofficial world record, a pleased SSW management gave Müller a bonus of 1,500 Marks.

On October 7, 1917 *Idflieg* ordered nine fighter prototypes (numbered D.7550/17 to 7558/17) to explore the possibilities of this new combination.

These had lighter, stronger airframes specifically designed for the new Sh.III engine. These prototypes were continually modified (wing span, airfoil type, tail surfaces, etc., were all changed) in the quest for better performance and handling characteristics, as shown in the table on the facing page.

SSW Fighters at the First Fighter Competition

On December 26 1917, *Idflieg* ordered a pre-production batch of 20 D.III fighters (D.8340–8359/17), at the time the customary number for front-line evaluation and, in this case, for the First Fighter Competition. Entered in the Competition, which started January 20, 1918, were SSW prototype fighters D.7551/17, 7552/17, 7553/17 (ready January 24) and series D.III 8340/17. Made possible by the Sh.III engine, SSW pilot Hans Müller demonstrated the fighters' impressive performance and maneuverability. As an example, on January 21 D.7552/17 climbed to 6,000 meters in 21.5 minutes.

Regardless of their exceptional climb performance, the new SSW fighters were not yet perfected. In particular, they had challenging handling qualities, especially on landing, and required expert piloting.

SSW Fighter Prototypes Ordered October 7, 1917

Type	Serial #	Notes
D.IIc *kurz*	D.7550/17	8.5 m span, first flight October 22, 1917. Later redesignated D.III 7550/17 and became the static test airframe for the D.III production series.
D.IIc *lang*	D. 7551/17	9.0 m span, first flight November 15, 1917. Redesignated D.III, it became the true prototype of the D.III series. It participated in the First Fighter Competition. After being damaged, it was rebuilt as D.IV 7554/17 with a span of 8.35 m. Again damaged, it was rebuilt as D.IVa 7554/17 with a span of 7.4 m. It was accepted on May 4 and participated in the Second Fighter Competition in June.
D.III	D.7552/17	Made its first flight December 20, 1917, and participated in the First Fighter Competition. 8.7 m span, wing area 21.6 m^2, 6 m long, 500 kg empty weight.
D.IIe	D.7553/17	It was fitted with a wireless wing cellule with duralumin wing spars. During test flights the wings flexed too much and full wire bracing had to be fitted. Participated in the First Fighter Competition. Later converted to D.IV 7553/17 and delivered to *JG* II in April for evaluation in combat. Engine problems required its return to the factory, where it was modified and fitted with a new engine. It was returned to *JG* II in July 1918. Original span 8.2 m, length 6 m, 500 kg empty weight.
D.IV	D.7554/17	Was rebuilt from 7551/17; see above.
D.IV	D.7555/17	Prototype for the D.IV, made its first flight June 18, 1918. Delivered to Adlershof in August as the D.IV static test airframe. Span 8.2 m.
D.V	D.7556/17	D.IV fuselage with 2-bay wireless wings, duralumin spars. First flew June 14, 1918.
D.V	D.7557/17	Participated in the Second Fighter Competition.
D.V	D.7558/17	Slated for home defense like the other D.V fighters but the Armistice intervened.

Front-line pilots, accustomed to more docile fighters, experienced difficulty flying the SSW fighters.

On January 23 Müller took 7553/17 up for its first flight. The aircraft proved extremely nose heavy and required full elevator and throttle to maintain level flight. Coming in to land, the engine failed at 1,000 meters. In spite of full elevator, the nose-heavy fighter dived toward the ground. At the very last moment Müller regained control, touched down sharply, and buried the nose in the sand with only minor damage! Director von Siemens, impressed by Müller's flying ability, was of the opinion that if the accident had occurred at the small SSW airfield rather than huge Adlershof, the fighter would have

Right: The fighters powered by the advanced rotary engines had much better climb rates than the in-line engine fighters due to the light weight and high power-to-weight ratio of their rotary engines. However, at this time none of these advanced rotaries were in production and only the Sh.III eventually entered production in time to power fighters in combat. The Sh.III was typically rated at 160 hp, the power developed by the prototype engines, but production engines produced 205 hp.

Climb Rates to 5,000 m at the First Fighter Competition

Type	In-Line Engines	Minutes	Notes
Albatros D.Va	Mercedes D.III	36.3	Production aircraft
Albatros D.Va	Mercedes D.IIIa	25.5	Production aircraft
Albatros D.Va	BMW D.IIIa	18.5	Pre-production engine
Fokker V11	Mercedes D.III	25.2	
Fokker V18	Mercedes D.III	28.0	
Pfalz D.IIIa	Mercedes D.III	33.0	Production aircraft
Rumpler D.I	Mercedes D.III	23.8	
	Rotary Engines		
Fokker Dr.I	Goebel Goe.III	14.0	Experimental engine
Fokker Dr.I	Oberursel Ur.III	15.5	Experimental engine
Pfalz D.VIII	Sh.III	13.8	Pre-production engine
SSW D.III	Sh.III	13.0	8340/17

Siemens-Schuckert D-Type Fighter Specifications					
	D.I	**D.III**	**D.IV**	**D.V**	**D.VI**
Engine	110 hp Sh.I	205 hp Sh.III	205 hp Sh.IIIa	205 hp Sh.III	205 hp Sh.IIIa
Armament	One, later two guns	Two guns	Two guns	Two guns	Two guns*
Span, Upper	7.50 m	8.40 m	8.35 m	8.86 m	9.37 m
Span, Lower	6.30 m	8.40 m	8.35 m		na
Chord, Upper	1.30 m	1.46 m	1.0 m		
Chord, Lower	0.8 m	1.0 m	1.0 m		na
Length	6.00 m	5.85 m	5.58 m	5.7 m	6.50 m
Wing Area	14.4 m²	18.84 m²	15.12 m²		12.46 m²
Empty Weight	430 kg	534 kg	540 kg	514 kg	540 kg
Loaded Weight	675 kg	725 kg	735 kg	725 kg	710 kg
Max. Full Weight:	735 kg	740 kg	735 kg	734.5 kg	735 kg
Maximum Speed:	155 km/h	180 km/h	190 km/h		220 km/h
Climb: 1,000m	3.5 minutes	1.75 minutes	1.9 minutes		
2,000m	8.0 minutes	3.75 minutes	3.7 minutes		
3,000m	14.5 minutes	6.0 minutes	6.4 minutes		
4,000m	24.3 minutes	9.0 minutes	9.1 minutes		
5,000m	45 minutes	13.0 minutes	12.1 minutes	17.4 minutes	
6,000m		20.0 minutes	15.5 minutes		16 minutes
7,000m					22 minutes
Ceiling:		8,000m	8,000m		8,000m
Duration:	2.5 hours	2 hours	2 hours		2 hours

Notes:
1. Although the Sh.III and Sh.IIIa developed slightly over 205 hp (see pages 175–183), contemporary publications usually stated their power at 160 hp, the power developed by the initial prototype engines.
2. SSW D.III climb times from official German *Baubeschreibung* figures. SSW D.III with Sh.IIIa engine had these times to altitude: 5,000 meters in 10.5 minutes, 6,000 meters in 15 minutes, 7,000 meters in 22.5 minutes, and 8,000 meters in 34.5 minutes.
3. SSW D.IV climb times from G.P. Neumann figures.
4. *Lt.* Lenz recorded 14.5 minutes to 6,000m during an operational sortie in his SSW D.IV during which he shot down two British airplanes.
5. SSW D.VI armament is that projected for production; the two prototypes were unarmed.

been totally destroyed.

Also on January 23, *Oblt.* Bruno Loerzer, an experienced fighter ace, flew D.7552/17 for evaluation. He misjudged the landing ('the tail wouldn't come down'), had to go around the airfield again, and finally stalled in and flipped over, severely damaging the aircraft.

On January 24, Müller was taxiing the 'wireless' D.IIe 7553/17 when the undercarriage wire bracing failed, a failure attributed to substandard material.

On January 25, *Lt.* Hans von der Osten flew D.7551/17. Reporting unsuitably high

propeller revolutions, heavy control forces, poor maneuverability, and landing problems, Osten compared the SSW D.III to the Fokker D.VII prototype as 'an elephant to a mosquito'.

On January 26, after a high-altitude flight, *Lt.* Busse fared badly when D.7551/17 turned over upon landing in the sandy terrain, severely damaging the airframe and engine. By then it was obvious to everyone that the SSW D.III prototypes, with their touchy controls and tall undercarriages (required by the large two-bladed propeller), demanded a degree of flying skill beyond a typical fighter pilot's

Above: SSW D.IIc 7551/17 with 9.0 m span first flew Nov. 15, 1917. It was redesignated as a D.III and became the true prototype of the D.III production aircraft. It competed at the First Fighter Competition in January 1918.

ability. The landing problem was greatly improved by changing to a four-bladed propeller which, being of smaller diameter, permitted shortening the undercarriage. Overall handling qualities were improved by modifications to the ailerons and tail surfaces. With impressive speed, D.III D.8340/17, which had not yet flown, was given a new elevator and undercarriage and returned to the Competition on January 28.

The spectacular climb rate of the rotary-engined fighters powered by the Sh.III and Goebel Goe.III engines was accompanied by good speed. On January 23, various fighters piloted by company and front-line pilots went aloft for 'parallel competition'. Müller in SSW D.III 8340/17, and making the first flight in this first production D.III, competed with Manfred von Richthofen flying the Fokker D.VII prototype (V11 or V18) and *Lt*. Hans Klein in an Albatros D.Va powered by the new 180 hp Mercedes D.IIIaü high-compression engine. The SSW D.III was

Below: SSW D.7551/17, now designated a D.III, at the First Fighter Competition in January 1918.

SSW D.III & D.IV Production Orders

Order Date	Siemens #	Quantity	Military #	Type
17 Jan. 1918	11054	20	D8340 – 8359/17	D.III
1 March 1918	11080	30	D1600 – 1629/18	D.III
22 Apr. 1918	11111	30	D3007 – 3026/18 D3037 – 3046/18	D.III D.III
22 Apr. 1918	11122	20	D3027 – 3036/18 D3047 – 3056/18	D.IV D.IV
7 May 1918	11115	50	D3060 – 3109/18	D.IV
26 July 1918	11163	60	D6150 – 6209/18	D.IV
10 Sep. 1918	FW38	50	D9000 – 9049/18	D.IV
16 Oct. 1918	FW90	100	D11500 – 11599/18	D.IV
End Oct. 1918	FW92?	100	?	D.IV

In addition to the three D.III prototypes and three D.IV prototypes, Siemens records show a total of 80 production D.III aircraft and 280 production D.IV aircraft were ordered as shown in the table above. Another order for 100 D.IV at the end of October is not listed in the Siemens records but may be Siemens order FW92, for which no other details were recorded. The total number actually delivered is not known, but it is very unlikely all the D.IV aircraft ordered were delivered. There are small discrepancies between order dates given above and those listed in the Siemens records, possibly caused by change orders.

substantially faster than the Albatros and marginally faster than the Fokker below 2,000 meters.

With most of the attention going to the prototype Fokker D.VII, which was declared the winner of the competition, SSW Director von Siemens thought the SSW prototypes did not receive appropriate recognition for their performance and opined, "one has the feeling that Richthofen's sympathies, as before, lie with Fokker!"

During his inspection of the Siemens factory on January 24, Richthofen stated, "all aircraft at the Competition were too slow; the rate of climb was of secondary importance." This opinion was shared by Loerzer and the other combat pilots. SSW responded by proposing the SSW D.IV that, given a smaller upper wing chord and reduced wing area, sacrificed climb performance for increased speed. Interestingly, the performance specifications show

Below: SSW D.III 7551/17 at the First Fighter Competition in January 1918. All control surfaces are aerodynamically balanced, but this early stage of development the wings did not yet have the horn-type aerodynamic balances; instead, these were inset into the wing profile.

Above: SSW D.III 7551/17 at the First Fighter Competition in January 1918. At this early stage of development the engine has a full cowling and there are no cooling louvers in the spinner.

Below: SSW D.III 8340/17, the first production D.III fighter, is shown after assignment to *Vzfw*. Fritz Beckhardt of *Kest* 5. Beckhardt, who was Jewish, chose the reversed swastika, a traditional Nordic symbol of good luck, as his personal marking. This had no political significance during WWI and was used on some Allied aircraft as well. The swastika only became controversial after its use by the National Socialist German Worker's Party starting in the 1920s.

Above: SSW D.III 8340/17, the first production D.III fighter of *Kest* 5 is shown after *Vzfw*. Fritz Beckhardt's crash landing at Gossau, near Rapperswill, Switzerland, November 13, 1918. A number of German pilots flew to Switzerland rather than surrender their aircraft to the Allies, including four SSW D.III fighters from *Kest* 5, suggesting prior agreement to do so.

the D.IV as having better climb rate than the D.III at high altitudes, indicating individual variations in conditions and piloting technique were greater than inherent design differences for climb performance.

On March 1 1918, *Idflieg*, on the basis of the competition results, ordered 30 D.III fighters (1600–1629/18) followed by 50 D.III (3007–3056/18) on March 23 and 50 D.IV (3060–3109/18) on April 8. Due to the priority of the faster D.IV, on June 8 *Idflieg* changed the March 23 order to 30 D.III (3007–3026/18 & 3037–3046/18) and 20 D.IV (3027–3036/18) fighters.

Initial Combat Experience

From March 16 to May 18, SSW shipped 41 D.III fighters and one D.IV to the front, with most going to *Jagdgeschwader* II (See Table below). *Hptm.* Rudolf Berthold, CO of *JG* II, during a squadron visit by SSW designer Bruno Steffen on April 22 1918, praised the D.III's 'brilliant' rate of climb, but requested increased speed, maneuverability, and airframe strength. The D.III and Sh.III engine combination was considered 'faultless' and had gained the 'trust of the pilots'. Maintenance of engine power at higher altitude was appreciated. The sole D.IV was reported 'in all respects superior to the

SSW Fighter Deliveries in Spring 1918			
Date	**Unit**	**Qty.**	**Serial Numbers**
16 March	*JG* III	6	D.III 8340–8345/17
6 April	*JG* II	9	D.III 8346–8354/17
15 April	*JG* II	1	D.IV 7553/17
19 April	*JG* II	10	D.III 8355–8359/17, 1600–1603 & 1605/18
30 April	*JG* II	6	D.III 1604, 1606–1608, 1611, 1612/18
13 May	*JG* II	6	D.III 1614–1619/18
18 May	*JG* II	4	D.III 1610, 1613, 1621, 1624/18

Left: From March 16 to May 18, 1918, SSW shipped one D.IV fighter and 41 D.III fighters to operational fighter units on the Western Front. As summarized in the table, all but six were delivered to *JG* II, then led by *Hptm.* Rudolf Berthold.

Above & Below: SSW D.III 8341/17, the second production D.III fighter, is shown at the SSW factory at Siemensstadt. Normally fitted with a spinner, the photo below shows it without its spinner.

D.III'.

Prohibited from flying over the front, *JG* II pilots intercepted and shot down two high-flying Breguet 14 bombers. But after seven to ten hours, the engines began to fail. They overheated, pistons seized, and piston heads disintegrated, falling into the crankcase. Even the Le Rhone rotary, which had a good reputation for reliability, showed similar symptoms when tested with Voltol, a synthetic castor-oil replacement. Investigation revealed that

This Page: SSW D.III 8341/17, the second production D.III fighter is shown at the SSW factory at Siemensstadt. Its full cowling and camouflage printed fabric on the wings and movable tail surfaces are clearly shown.

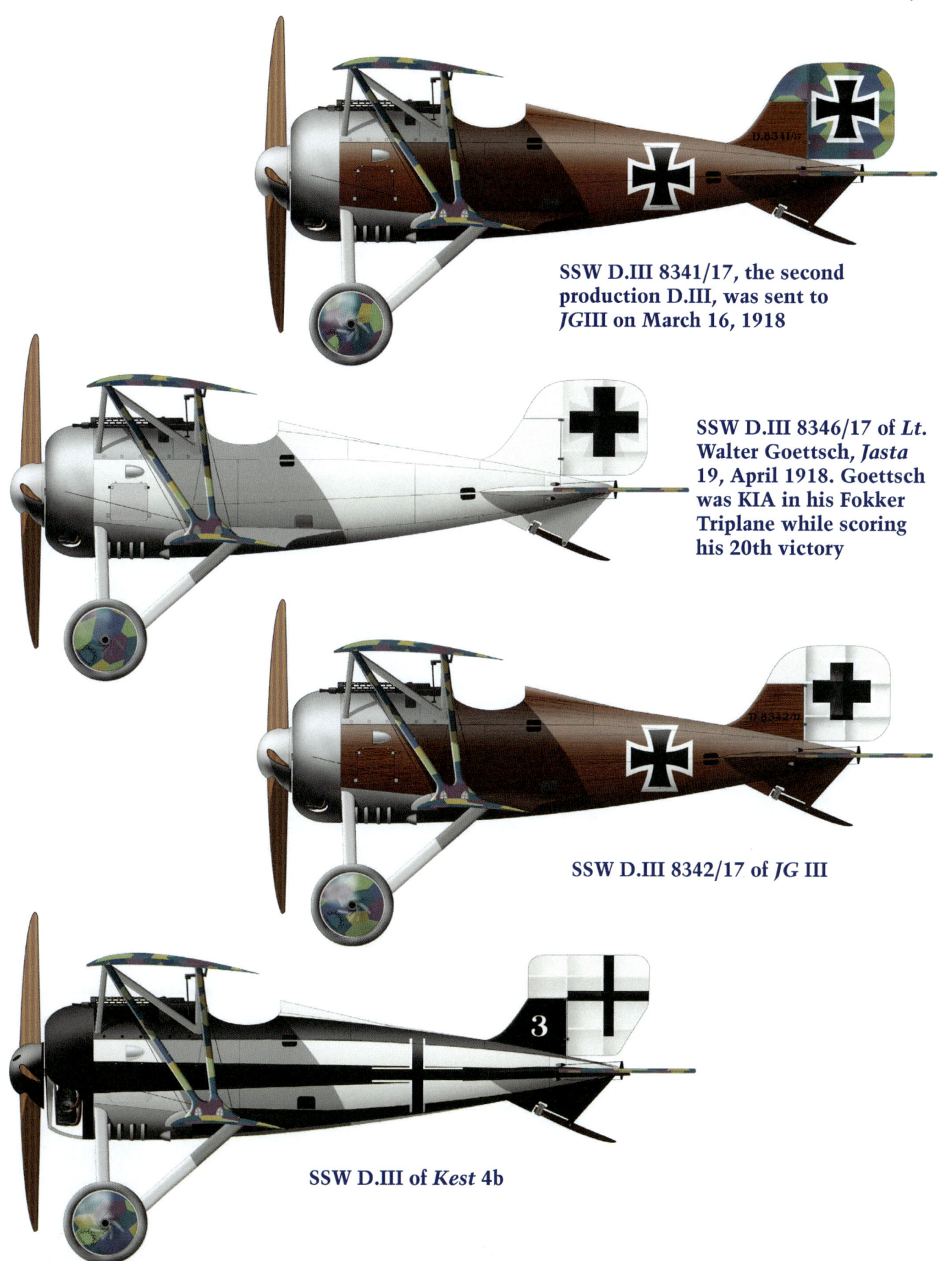

SSW D.III 8341/17, the second production D.III, was sent to *JG*III on March 16, 1918

SSW D.III 8346/17 of *Lt.* Walter Goettsch, *Jasta* 19, April 1918. Goettsch was KIA in his Fokker Triplane while scoring his 20th victory

SSW D.III 8342/17 of *JG* III

SSW D.III of *Kest* 4b

wrong viscosity oil had been delivered in incorrectly-marked barrels. On May 23 Berthold urged "the Siemens fighter be made available again for front-line use as quickly as possible for, after elimination of the present faults, it is likely to become one of our most useful fighter aircraft."

All the crippled SSW fighters were returned to the factory in May 1918 for engine replacement and extensive modifications. On June 19 *Idflieg* required that the existing D.III fighters and those in production receive the following major changes: a new rudder shape, rounded (not pointed) aileron balances (and rounded elevator balances on future production machines), shortened wing span, larger cockpit opening, new propeller pitch to maximize speed, a cut-away cowling, and a smaller propeller spinner.

Operational use of the Sh.III engine was prohibited pending successful completion of the demanding 40-hour endurance test, which was not achieved until July. Although the early-production Sh.III engines were unreliable, the revised Sh.IIIa, the version that passed the endurance testing and also known as the Sh.III *neu*, was a reliable engine. Furthermore, the Sh.III(Rh) engines built under license by the Rhenania Motorenfabrik AG (Rhemag) were also reliable. Eventually, about 40% of SSW fighters were powered by Rhemag-built engines. The availability of Rhemag-built engines coupled with the Siemens-Halske Sh.IIIa engine passing its endurance test permitted deliveries to the front (five D.III and one D.IV) to re-start on July 22 after a two-month interruption.

Above & Below: SSW D.III 8342/17, the third production D.III fighter is shown at the SSW factory at Siemensstadt. Like 8341/17 it had a full cowling and camouflage printed fabric on the wings and movable tail surfaces.

Above & Below: SSW D.III 8342/17, the third production D.III fighter is shown at the SSW factory at Siemensstadt. Like other early-production D.III fighters it had a full cowling with large spinner and camouflage printed fabric on the wings and movable tail surfaces. Factory personnel pose with the fighter above; below is ace and SSW test pilot Hans Müller.

Above: SSW D.III 8342/17, the third production D.III fighter is shown apparently after it reached JG III. The rudder has now been painted white and a *Balkenkruez* applied; the fuselage still retains the original *Eiserneskreuz*.

SSW D.V

Aircraft D.7556–7558/17 were the SSW D.V prototypes; these aircraft were ordered along with the D.III and D.IV prototypes on October 7, 1917. D.V. 7556/17 performed its first flight on June 14, 1918 and D.V 7557/17 participated in the Second Fighter Competition. Using essentially a D.IV fuselage, the D.V differed in being fitted with duralumin wing spars and 'wireless' two-bay wings. The three completed D.V. fighters were slated for home defense service but the Armistice intervened. Unfortunately, no photographs of these aircraft have surfaced.

SSW at the Second Fighter Competition

From the perspective of SSW, the Second Fighter Competition served primarily to compare production SSW fighters with the latest German prototypes. Eight SSW fighters, some powered by the Sh.IIIa engine and many fitted with the above modifications, were entered in the Second Fighter Competition that was held from May 27 to June 28 1918. Siemens reported that, 'because of significant weight increase and reduced wing area, the climb rates were somewhat reduced: 6,000 meters in 25–28 minutes.' With respect to speed and maneuverability,

the aircraft, 'showed up very favorably against the competition'. In the front-line pilots' de-briefing sessions (July 6 and 14), among the Sh.III-powered fighters the SSW D.IV, fitted with four ailerons, was chosen over the less-maneuverable Pfalz D.VIII, the pilots' second choice, which was recommended for home-defense duties. Accordingly, on July 26, *Idflieg* ordered 60 SSW D.IV fighters (6105–6209/18), followed by 50 fighters (9000–9049/18) on September 10, 100 (11500–11599/18) on October 16, and 100 at the end of October 1918.

The SSW D.III & D.IV in Action

SSW delivery records show that from July through November 1918, at least 136 SSW D.III and D.IV fighters were delivered to combat units. In general the D.III, due to its potentially higher climb rate, was supplied to the home-defense squadrons (*Kampfeinsitzer Staffeln* — *Kests* 2, 4a, 4b, 5, 6 and 8) and the faster D.IV to the Western Front: *JG* II, *Jasta* 14, *Jasta* 22, *Jastaschule* 1, and *Marine Jagdgruppe*. The few available combat reports universally praised the Siemens fighters but not without some criticism, especially from units having flown the more benign Fokker D.VII. Both the D.III and D.IV needed more flying skill for they were

Right: SSW D.III 8345/17 of the first D.III production batch in service at *JG* III in the spring of 1918. The backing plate for the large spinner is visible, showing all the cooling holes in it, and the aircraft may have been flown this way due to cooling problems with the new Sh.III counter-rotary.

Right & Below: The SSW D.III was powered by the same 205 hp Siemens-Halske Sh.III used in the Pfalz D.VIII; the SSW D.III had similar performance but was more maneuverable than the Pfalz D.VIII. These photos show early production SSW D.III 8346/17 from the first production batch assigned to *Lt*. Walter Goettsch, then *Staffelführer* of *Jasta* 19. It has the large spinner, full cowling, and inset aerodynamic balances of the first D.III production batch. Fokker Triplanes are lined up in the background.

Above & Below: Two more views of early production SSW D.III 8346/17 from the first production batch assigned to *Lt.* Walter Goettsch of *Jasta* 19. The fuselage has been painted white, Goettsch's personal color, and the propeller blades are protected by canvas coverings. Unfortunately, while flying his Fokker Dr.I on April 10, 1918, Goettsch was killed in the combat in which he scored his 20th victory. He may never have had the opportunity to fly his SSW. D.III in combat.

exceptionally maneuverable, very sensitive on the controls, and, unlike the docile Fokker D.VII, spun with little warning.

Therefore Müller and *Lt.* Bruno Rodschinka, both SSW test pilots, visited operational units to instruct on the SSW fighters and gave flying demonstrations in them that were generally considered the finest exhibition of low-altitude aerobatics ever seen by service pilots. Their efforts produced the necessary confidence in the types, as did the initial victories. Naval *Lt.z.S.* Theo Osterkamp shot down a D.H.4 at 6,000 meters on August 21 while Müller was

Left & Below: An early-production D.III at *Jasta* 19. The full image at left shows a triplane undergoing maintenance. The enlargement below shows another SSW D.III in the background.

Right: An early-production D.III at *Jasta* 19 undergoing engine maintenance with spinner and cowling removed. A *Jasta* 19 Fokker Dr.I is at left.

present. At *Kest* 8 Rodschinka downed two D.H.4 bombers out of 24 on September 7. A pilot of *Kest* 8 wrote on October 2, after two British aircraft were downed, that, 'we all now have the Siemens and are very satisfied… the Siemens is much superior to Allied aircraft and the Sh.III engine operated without complaint.' Even *Jagdgeschwader Richthofen*, who had been adverse to the Siemens fighters, after a demonstration by Hans Müller on October 5 requested 'twelve good D.IV fighters and a further twelve as soon as the first are shipped.'

Lt. Lenz, CO of *Jasta* 22, reported on October 3 that the D.IV was, 'superior to all aircraft at the front in climb, maneuverability, and speed above 4,000 meters'. At that altitude, 'it was impossible to fly formation with the Mercedes-engined Fokker D.VII'. Pilot training was mandatory since many were unfamiliar with the SSW fighters' rotary engine and their high landing speed. Above 5,000 meters the

high wing loading became noticeable in turns and required, 'watchfulness and opposite rudder'. Spin recovery was quick and effortless. The pilot's field of view was quite good, except straight down due to the round fuselage.

Although Germany had difficulty finding sufficient aircraft to meet the Armistice stipulations, it appears SSW fighters were held back because only a few were turned over to the Allies. Of the 44 D.IV fighters completed by SSW after the war, from December 1918 through July 1919, some were sent to *Flieger Abteilung* 431 of the *Grenzschutz Ost*, but most were placed in storage at *Reichswehr* depots at Johannisthal, Liegnitz, Thorn, and Döberitz before being destroyed in accordance with the Treaty of Versailles. At least one of the D.IVs shipped to Thorn later turned up in the hands of *Flieger Abteilung* 424 in Lithuania in 1919 (see photos page 2).

Above & Below: Two views of early production SSW D.III 8349/17 from the first production batch assigned to *Kest* 4b, photographed postwar in French hands. Tactical number '9' is painted on the fin in the manner of *Kest* 4b and the pilot's individual marking is painted on both sides of the fuselage. This aircraft has been rebuilt, accounting for the cut-away engine cowling, spinner cooling louvers, horn-balanced ailerons, and enlarged rudder for better maneuverability.

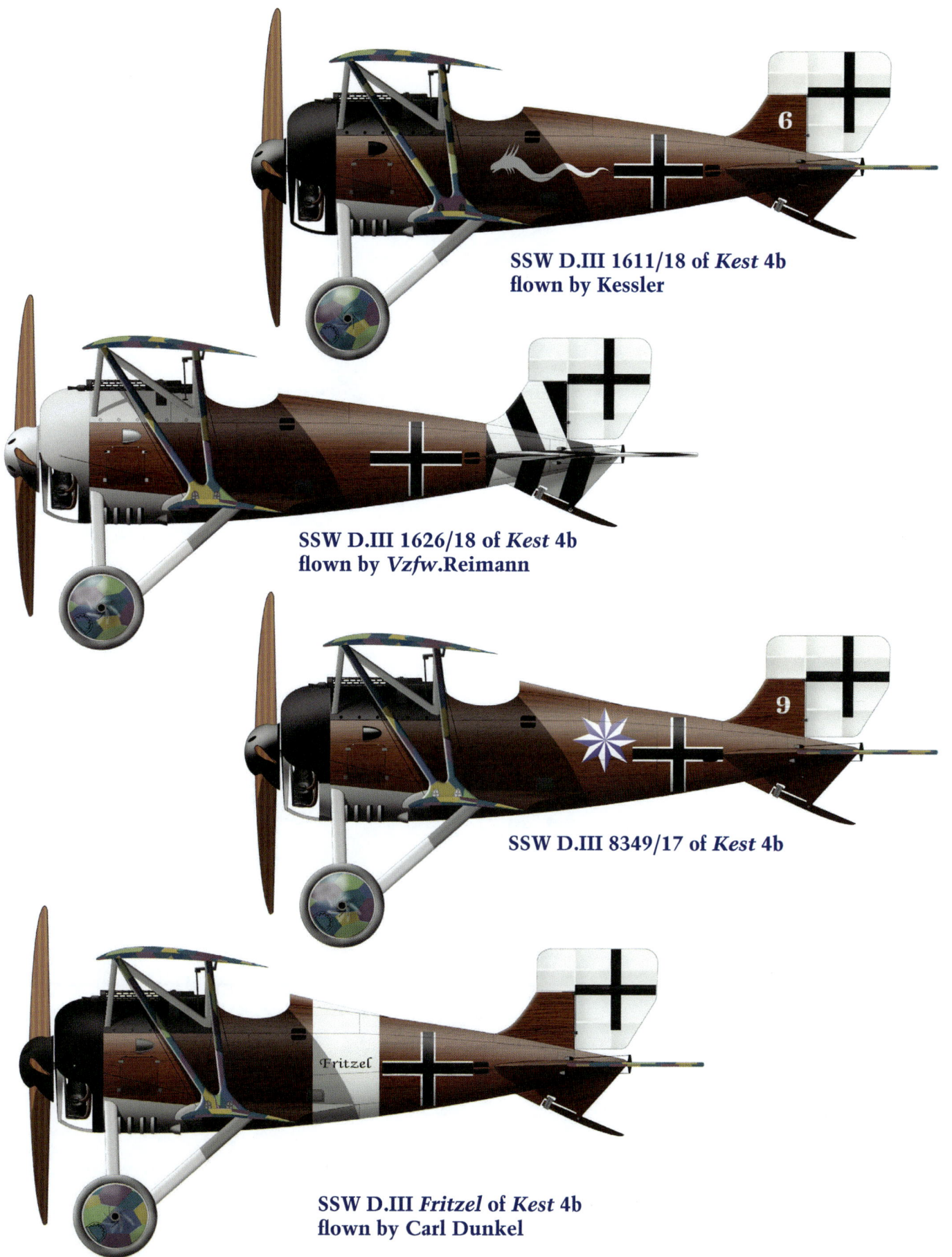

SSW D.III 1611/18 of *Kest* 4b
flown by Kessler

SSW D.III 1626/18 of *Kest* 4b
flown by *Vzfw.*Reimann

SSW D.III 8349/17 of *Kest* 4b

SSW D.III *Fritzel* of *Kest* 4b
flown by Carl Dunkel

Above & Below: Two views of early production SSW D.III 8344/17, the fifth production D.III, after rebuilding, accounting for the cut-away engine cowling, spinner cooling louvers, horn-balanced ailerons, and enlarged rudder. Assigned to *Kest* 5, after the Armistice the pilot, *Offz.* Arnold Eger, flew this aircraft to Switzerland, and it is seen after Swiss markings were applied; note the differences in rudder markings. The pilot's individual marking of skull and crossbones is painted on both sides of the fuselage.

Above Left: *Gefr*. Bruno Lange with SSW D.III 8344/17, probably at *Kest* 5's field in Lahr-Dinglingen.

Above Right: *Lt*. Eugen Weber with his SSW D.III 8357/17 at *Kest* 4b.

Below: Rebuilt first production series SSW D.III, almost certainly 8350/17, assigned to Ernst Udet at Metz-Frescaty airfield., Oct. 1918. Rebuilt D.III aircraft were distinguished by a cutaway cowling and louvers in the spinner for improved cooling. In addition, they had horn-balanced ailerons and a larger rudder for improved maneuverability and handling qualities.

SSW D.IIIs of *Kest* 5

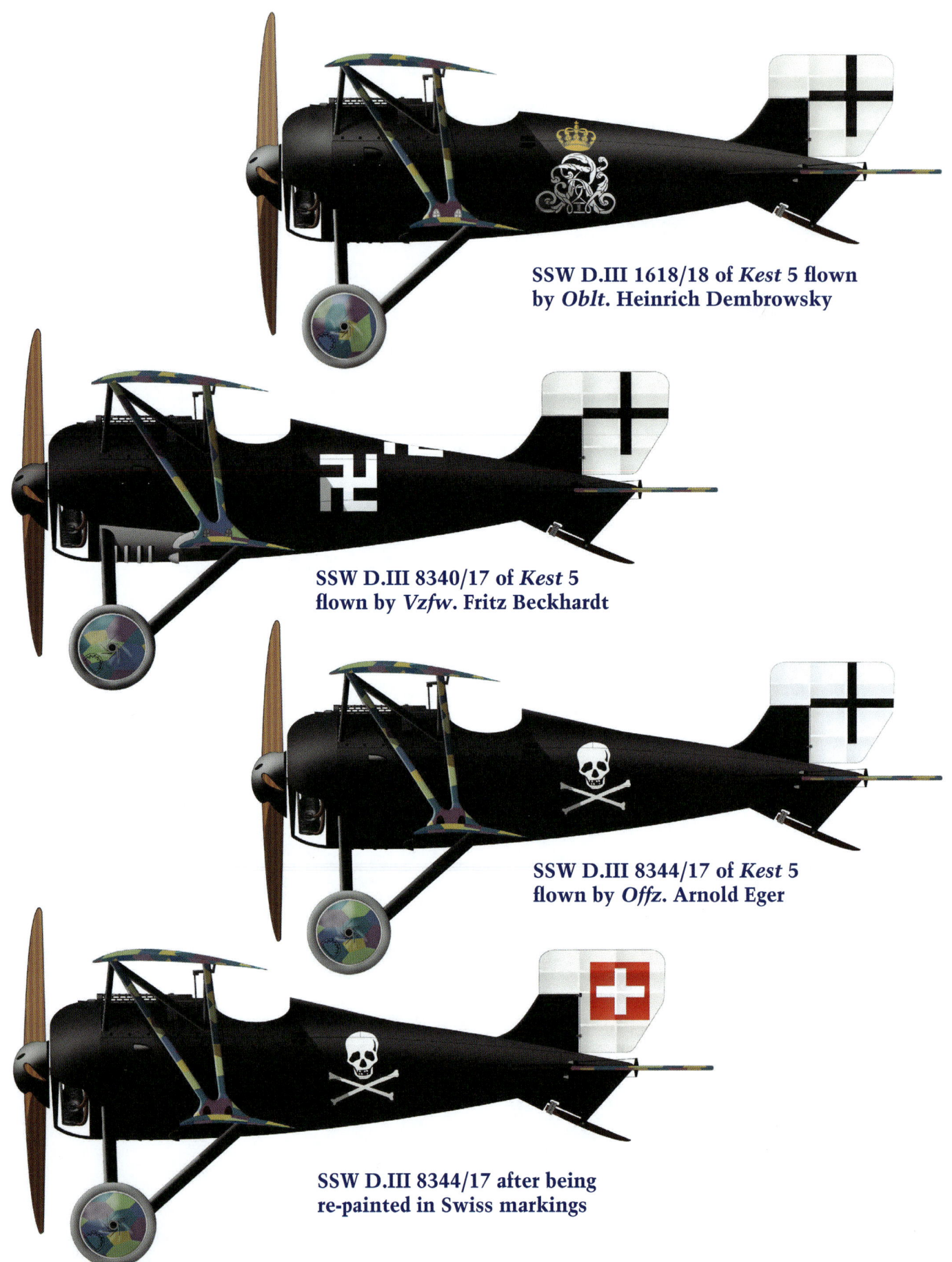

SSW D.III 1618/18 of *Kest* 5 flown
by *Oblt.* Heinrich Dembrowsky

SSW D.III 8340/17 of *Kest* 5
flown by *Vzfw.* Fritz Beckhardt

SSW D.III 8344/17 of *Kest* 5
flown by *Offz.* Arnold Eger

SSW D.III 8344/17 after being
re-painted in Swiss markings

Above: Carl Dunkel and his SSW D.III *Fritzel* of *Kest* 4b. The serial is unknown but the headrest shows it was an early production aircraft that has been re-built with revised flight controls and cooling modifications.

Below: Kessler and his SSW D.III 1611/18 of *Kest* 4b. The aircraft was built in the second production batch and wears Kessler's personal marking on the fuselage and the tactical number "6" on the fin that was typical of *Kest* 4b practice.

Above: Lineup of *Kest* 4b with Kessler's SSW D.III second from right.

Left: SSW D.III postwar with markings suggesting it was from *Kest* 4b.

Below Left: SSW D.III first series cockpit, probably 8341/17 at Siemensstadt.

Below: SSW D.III with non-standard propeller made from two 2-blade propellers, probably post-Armistice.

Above: SSW D.III 1618/18 of *Kest* 5. The pilot's personal marking is the monogram of Frederick the Great.

Below: SSW D.III 1618/18 after a crash. The serial on the wheel cover is 8346, but this is not a not a first series D.III because it has no headrest. Per discussions on The Aerodrome Forum, this was identified as D.1618/18, crash-landed at Schaffhausen, Switzerland, by *Oblt.* Heinrich Dembrowsky on November 13, 1918. The fuselage marking is based on the insignia of *Grenadier-Regiment King Friedrich Wilhelm I (2. East Prussia) Nr. 3*. Note open access covers for attaching lower wing at the spars.

Above: *Vzfw*. Reimann and his SSW D.III 1626/18 of *Kest* 4b; the stripes covered the horizontal tail and upper wing center.
Below: *Vzfw*. Paul Leim in front of his SSW D.III 1628/18 of *Kest* 4b.

Above: SSW D.III 1620/18 at Siemensstadt, showing final form of cowling, spinner, ailerons, and rudder. Wing and rudder crosses have been over-painted in 6:5 proportions.

Below: SSW D.III 1620/18 at Siemensstadt.

Left: *Vzfw.* Paul Leim in front of his SSW D.III 1628/18 of *Kest* 4b. The black and white fuselage bands were his personal marking in the *Kest*.

Right: SSW D.III 8342/17, the third production D.III fighter is shown at the SSW factory at Siemensstadt. Like other early-production D.III fighters it had a full cowling with large spinner, camouflage printed fabric on the wings and movable tail surfaces, and inset aileron balances. First series aircraft had a small headrest behind the cockpit that was eliminated in later production batches.

Below: SSW D.III 3025/18 with an American pilot in the cockpit postwar.

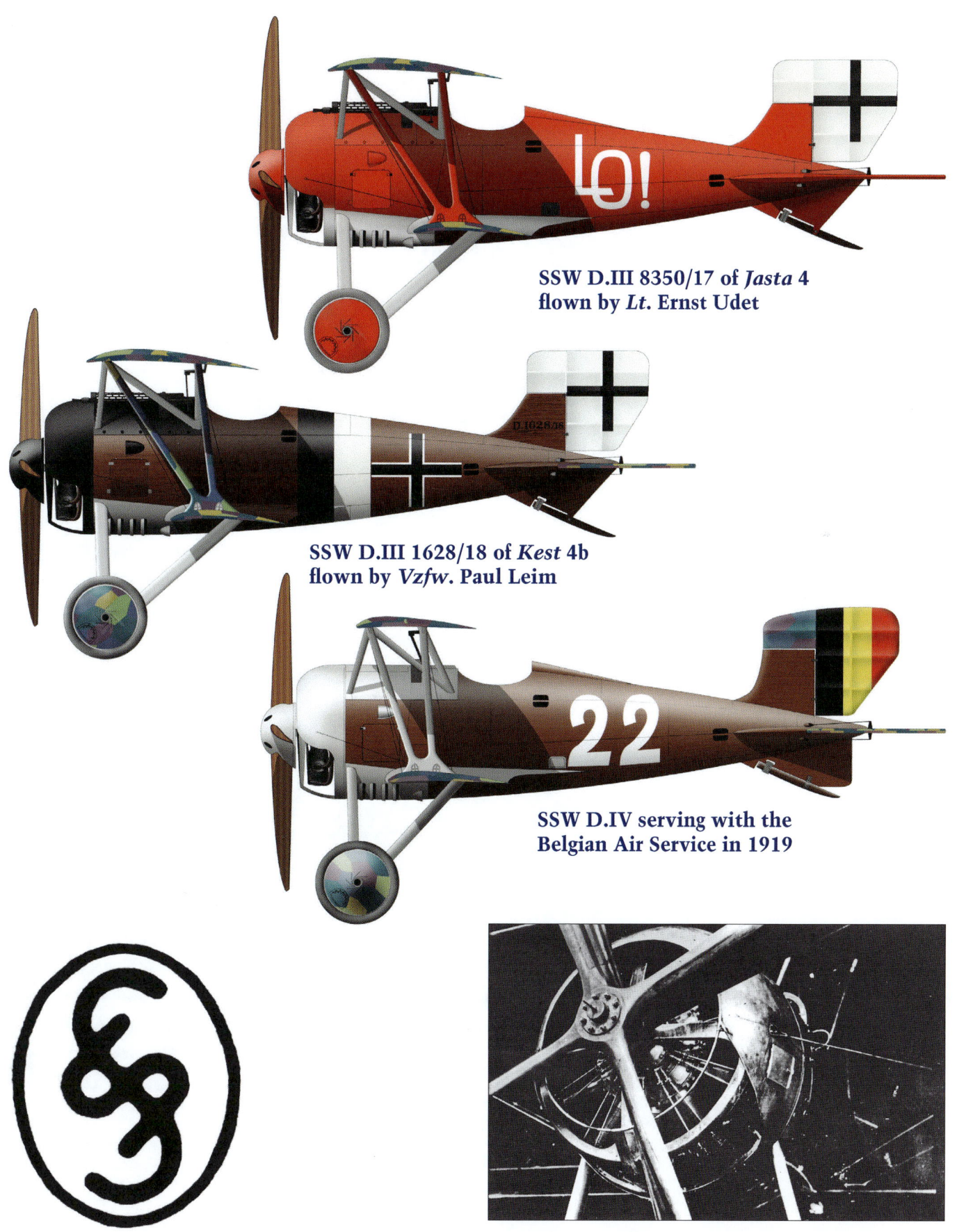

SSW D.III 8350/17 of *Jasta* 4
flown by *Lt.* Ernst Udet

SSW D.III 1628/18 of *Kest* 4b
flown by *Vzfw.* Paul Leim

SSW D.IV serving with the
Belgian Air Service in 1919

Above: SSW D.III 3008/18 with transitional cowling.

Above: SSW D.III 3008/18 of the third production batch. The cowling has been cut away for better cooling but there are no louvers in the spinner. Horn balances are now a feature of all four ailerons.

Below: SSW D.III 3025/18 of the third production batch is shown in American hands postwar. It now has the final cowling design and cooling louvers in the spinner. The unknown pilot had a white arrow on the fuselage for his personal marking.

Above & Below: SSW D.III 3025/18 of the third production batch is shown in American hands postwar. In addition to the final cowling design and cooling louvers in the spinner, it has the final control configuration with horn balances for all four ailerons. The engine is being run up in the photo above. D.III 3025/18 had been with *Kest* 8 in Bitsch.

This Page: SSW D.III 3025/18 of the third production batch is shown in American hands postwar. It is in the final production configuration.

Facing Page: SSW D.III 3025/18 with an American pilot in the cockpit postwar.

Above & Below: SSW D.III 3025/18 of the third production batch is shown in American hands postwar. In addition to the final cowling design and cooling louvers in the spinner, it has the final control configuration with horn balances for all four ailerons.

Above: *Lt.* Bruno Rodschinka, SSW test pilot, in front of SSW D.IV 6173/18 that was allocated to *Jasta* 14 on November 5, 1918. Together with *Lt.* Hans Müller, another SSW test pilot, Rodschinka toured units equipped with SSW fighters and gave stunning, low-level aerobatic displays to improve the combat pilots' opinion of the fighters.

Below: SSW D.III provided to *Lt.* Ernst Udet prior to painting his personal marking "LO!" on the fuselage.

SSW D.III

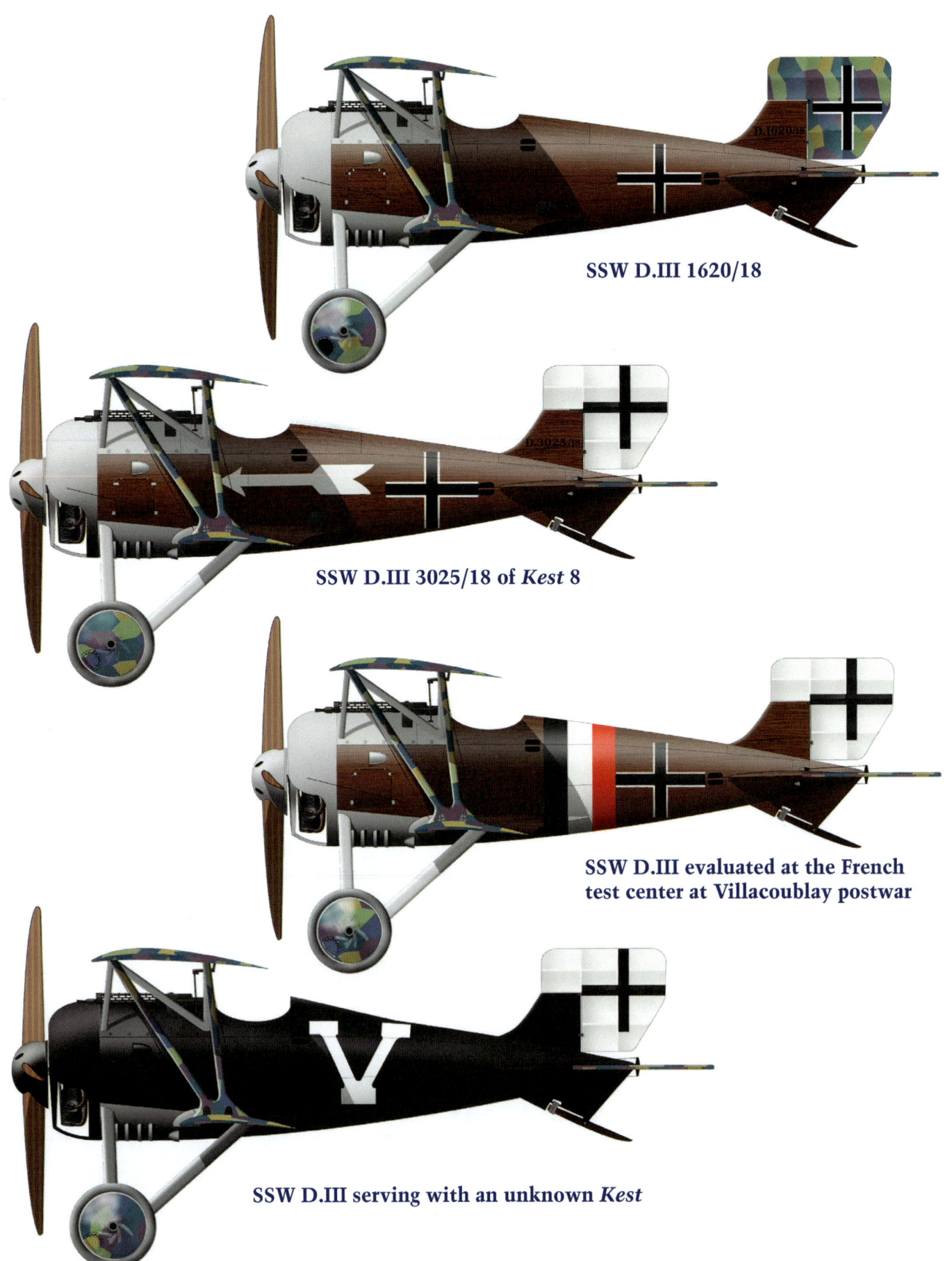

SSW D.III 1620/18

SSW D.III 3025/18 of *Kest* 8

SSW D.III evaluated at the French
test center at Villacoublay postwar

SSW D.III serving with an unknown *Kest*

Above & Below: SSW D.III probably serving with a *Kest*. It was a first series D.III, which means it could have been assigned to *Kest* 4b (Freiburg), *Kest* 5 (Lahr), *Kest* 6 (Bonn), or *Kest* 8 (Bitsch), but the *Kest* 4b lineup photo does not show it.

Above: SSW D.III of *Lt.* Joachim von Ziegesar of *Jasta* 15, May 1918. The three white leaves or feathers were the pilot's personal marking. Ziegesar scored three confirmed victories and was acting *Staffelführer* of *Jasta* 15 August 13–18, 1918.

Below: SSW D.III of *Lt.d.R.* Alfred Greven of *Jasta* 12. The cowling is not standard and there is no spinner. The white lightning bolt was Greven's personal marking. Greven scored two confirmed victories in September 1918 and two more in October for a total of four.

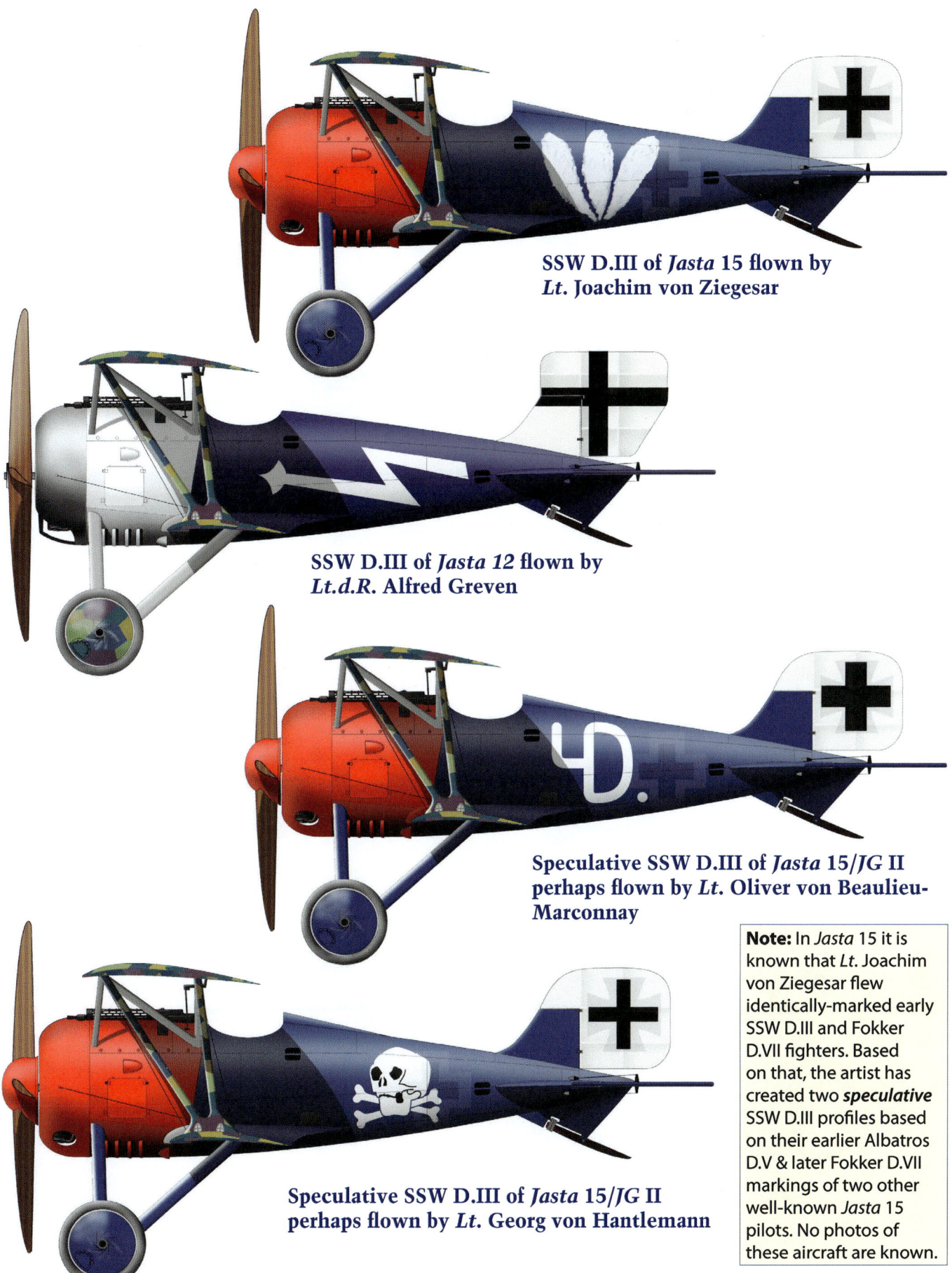

SSW D.III of *Jasta* 15 flown by
Lt. Joachim von Ziegesar

SSW D.III of *Jasta 12* flown by
Lt.d.R. Alfred Greven

Speculative SSW D.III of *Jasta* 15/*JG* II
perhaps flown by *Lt.* Oliver von Beaulieu-
Marconnay

Speculative SSW D.III of *Jasta* 15/*JG* II
perhaps flown by *Lt.* Georg von Hantlemann

Note: In *Jasta* 15 it is
known that *Lt.* Joachim
von Ziegesar flew
identically-marked early
SSW D.III and Fokker
D.VII fighters. Based
on that, the artist has
created two *speculative*
SSW D.III profiles based
on their earlier Albatros
D.V & later Fokker D.VII
markings of two other
well-known *Jasta* 15
pilots. No photos of
these aircraft are known.

SSW D.IV

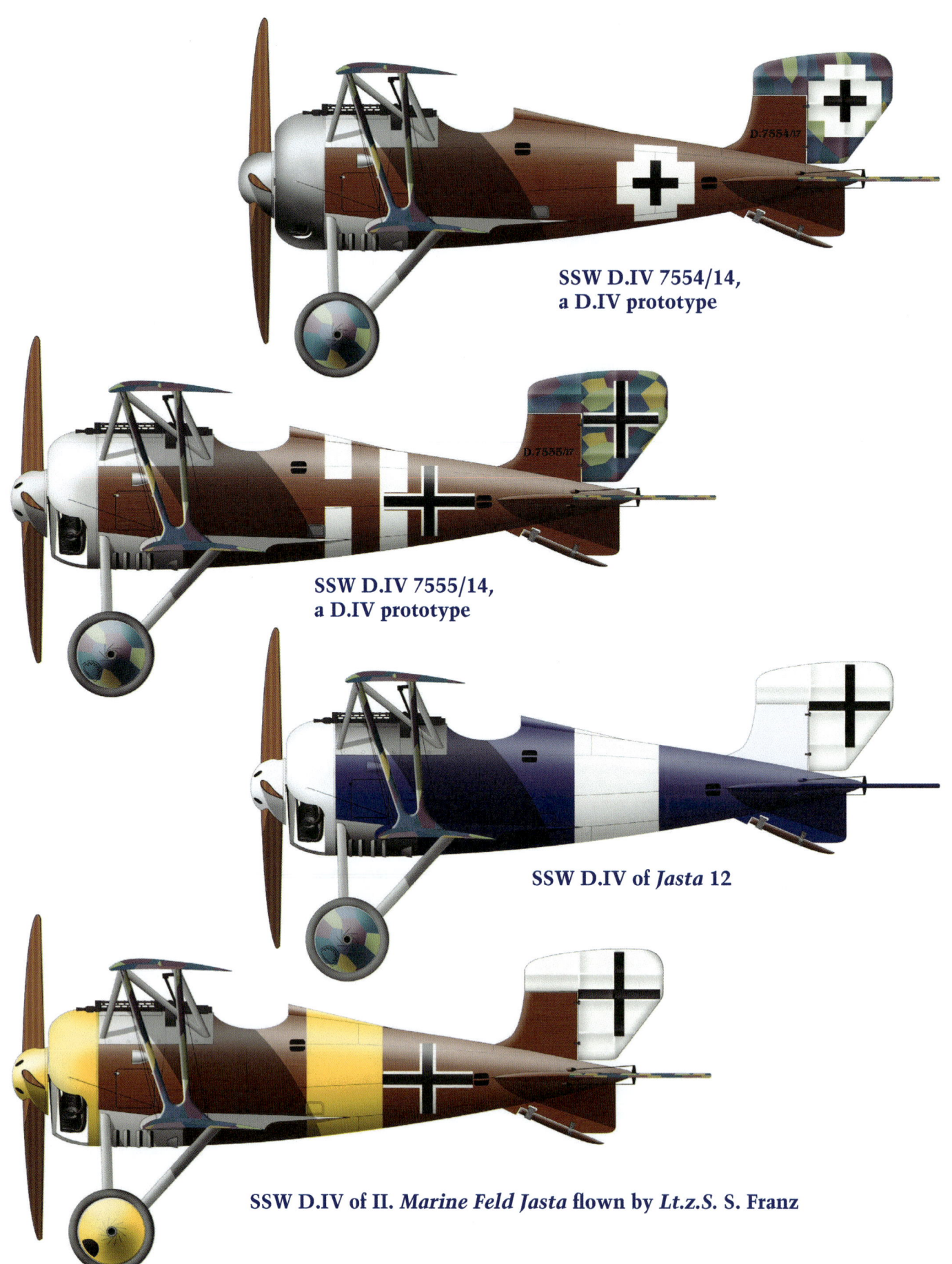

**SSW D.IV 7554/14,
a D.IV prototype**

**SSW D.IV 7555/14,
a D.IV prototype**

SSW D.IV of *Jasta* 12

SSW D.IV of II. *Marine Feld Jasta* flown by *Lt.z.S.* S. Franz

Above & Below: Late model SSW D.III under postwar evaluation at the French test center at Villacoublay.

Above & Below: Late model SSW D.III under postwar evaluation at the French test center at Villacoublay.

Below: Derelict SSW D.III (at right) and Spad (at left) enroute to be scrapped postwar.

Above: Lineup of SSW D.IV fighters of an unknown unit. The D.IV had a narrower-chord upper wing than the D.III, reducing wing area and drag and making it faster. The narrower upper wing reduced the distance between the spars, resulting in a smaller angle between the interplane struts, the key recognition feature distinguishing the D.IV from the D.III.

Below: A confident looking *Lt.z.S.* S. Franz stands in front of his SSW D.IV of the *II. MLF Jasta.*

Above: SSW D.IV 3028/18 of *MLF Jasta*, late August–early September, 1918.

Below: SSW D.IV 3048/18, *MLF Jasta,* August 30, 1918.

Right & Below: SSW D.IV 3049/18 crashed on its delivery flight while being flown by *Lt*. Speer.

Below: SSW D.IV 3082/18 of *JG* II photographed October 25, 1918.

Above & Below: SSW D.IV 3082/18 of *JG* II photographed October 25, 1918.

Above: SSW D.IV 7555/17 was the true prototype of the D.IV. It made its first flight June 18, 1918, and was delivered to Adlershof in August as the D.IV static test airframe.

Below: SSW D.IV 3082/18 of *JG* II photographed October 25, 1918.

Above & Below: *Lt.* Alfred Lenz, *Staffelführer* of *Jasta* 22, with his SSW D.IV 3083/18 photographed after August 1918. Lenz scored the last of his six confirmed victories with this aircraft on September 29, 1918 when he downed an S.E.5a. Lenz served at the front for 42 months, during which he logged more than 800 flying hours.

Above: *Lt.* Friedrich-Wilhelm Liebig of *Jasta* 22 in the cockpit of his SSW D.IV. *Lt.* Alfred Lenz, *Staffelführer* of *Jasta* 22, stands next to the cockpit.

Below: Apparently Liebig did not absorb all the advice given by Lenz; here is Liebig's SSW D.IV after he crashed it. Leibig scored one confirmed victory over a Sopwith Camel on October 4, 1918.

Above & Left: Crash of prototype SSW D.IV 7554/17. This aircraft was rebuilt from D.III 7551/17 after its crash.

Below: SSW D.IV 7555/17 was the true prototype of the D.IV. It made its first flight June 18, 1918. It was delivered to Adlershof in August as the D.IV static test airframe.

Above: A late model D.IV (with small spinner without louvers) with test gauge and other instruments attached to the fuel or oil system.

Below: Lineup of *Jasta* 12; SSW D.IV fighters at left and a Fokker D.VII at right.

Above: *Lt.z.S.* Bertram Heinrich of *MFJ I* posses with his SSW D.IV. Heinrich was an ace who scored 12 confirmed victories before being killed in action on August 31, 1918 while flying a Fokker D.VII.

Right: The cockpit of SSW D.III D.8341/17 in its original form. The SSW-designed control column was used only on first series machines. It was replaced on later and rebuilt D.IIIs with SSW's version of the Fokker *Steuerangriff* dictated by *Idflieg*.

Below: Wreck of D.6178/18, crashed by *Uffz.* Hassenmitter, *FA* 431, at Hundsfeld bei Breslau, 24 July 1919.

Above: SSW D.IV, probably postwar.

Left: Late model D.IV being inspected by civilians and officers, probably postwar.

Above: This SSW D.IV was found in dilapidated condition in Evere, Belgium in 1918 and completely refurbished by Belgian mechanics for *Lt*. Robin. The missing S-H blocktube carburetor was replaced by a Le Rhône unit. It competed in an early postwar air race.

Above: SSW D.IV in Belgian markings postwar.

Above: Belgian SSW D.IV at Evere in 1919.

Above: SSW D.IV in Belgian service postwar.

Above: SSW D.IV "22" in Belgian service postwar.

Above: SSW D.IV 3048/18.
Below: D.IV without tires, probably photographed postwar.

Above: SSW D.IV airframes ready for scrapping at Siemensstadt, 1920.

Above: Workers scrapping SSW D.IV airframes postwar.
Below: Lineup of SSW D.IV fighters.

Above: Closeup of SSW D.IV airframes.
Below: SSW D.III, possibly from a *Kest*, in French hands postwar.

Above: Late production SSW D.IV.

Left: Late model SSW D.IV in Sweden in 1920. The spinner has no cooling air scoops.

Albatros H.1

Above: The Albatros H.1, a SSW D.IV that was rebuilt postwar for high-altitude flight tests with larger wing and revised tail surfaces.

This Page & Facing Page: Details of the restored Albatros H.I, a modified SSW D.IV, in the museum at Krakow.

Right: *Erstes Höhenflugzeug* (First High-Altitude Aircraft) was painted on both sides of the aircraft as shown here prior to restoration.

Below Right: Front view of the Albatros H.1.

This Page: Details of the restored Albatros H.I, a SSW D.IV rebuilt with extended-span wings for high-altitude flight tests postwar, in the museum at Krakow.

Facing Page: Details of the restored Albatros H.I, a SSW D.IV rebuilt postwar for high-altitude flight tests, in the museum at Krakow.

Facing Page Upper Right: Details of the forward fuselage of an SSW D.IV, , showing MG ammo chutes, synchronizer cables and fuel tank with gauge and filler cap. The purpose of the angled fitting on upper right is unknown – it may be the fuel shutoff cock.

SSW D.III & D.IV Shipment Logs						
Type	**Series**	**Serial #**	**Order Date**	**Letter of Advice**	**Delivery Date**	**Consignee**
D.III	Prototype	D7550/17	7-Oct-17	D240Sfl	4-Jan-18	*Flugzeugmeisterei* Adlershof
D.III	Prototype	D7551/17	7-Oct-17	D268Sfl	18-Jan-18	*Flugzeugmeisterei* Adlershof
D.III	Prototype	D7552/17	7-Oct-17	D238Sfl	4-Jan-18	*Flugzeugmeisterei* Adlershof
D.III	1	D8340/17	17-Jan-18	D343Sfl	16-Mar-18	*JG* III, 7 *Armee*
D.III	1	D8341/17	17-Jan-18	D343Sfl	16-Mar-18	*JG* III, 7 *Armee*
D.III	1	D8342/17	17-Jan-18	D343Sfl	16-Mar-18	*JG* III, 7 *Armee*
D.III	1	D8343/17	17-Jan-18	D343Sfl	16-Mar-18	*JG* III, 7 *Armee*
D.III	1	D8344/17	17-Jan-18	D343Sfl	16-Mar-18	*JG* III, 7 *Armee*
D.III	1	D8345/17	17-Jan-18	D343Sfl	16-Mar-18	*JG* III, 7 *Armee*
D.III	1	D8346/17	17-Jan-18	D343Sfl	6-Apr-18	*JG* II
D.III	1	D8347/17	17-Jan-18	D343Sfl	6-Apr-18	*JG* II
D.III	1	D8348/17	17-Jan-18	D343Sfl	6-Apr-18	*JG* II
D.III	1	D8349/17	17-Jan-18	D343Sfl	6-Apr-18	*JG* II
D.III	1	D8350/17	17-Jan-18	D343Sfl	6-Apr-18	*JG* II
D.III	1	D8351/17	17-Jan-18	D343Sfl	6-Apr-18	*JG* II
D.III	1	D8352/17	17-Jan-18	D343Sfl	6-Apr-18	*JG* II
D.III	1	D8353/17	17-Jan-18	D343Sfl	6-Apr-18	*JG* II
D.III	1	D8354/17	17-Jan-18	D343Sfl	6-Apr-18	*JG* II
D.III	1	D8355/17	17-Jan-18	D343Sfl	19-Apr-18	*JG* II
D.III	1	D8356/17	17-Jan-18	D343Sfl	19-Apr-18	*JG* II
D.III	1	D8357/17	17-Jan-18	D343Sfl	19-Apr-18	*JG* II
D.III	1	D8358/17	17-Jan-18	D343Sfl	19-Apr-18	*JG* II
D.III	1	D8359/17	17-Jan-18	D343Sfl	19-Apr-18	*JG* II
D.III rebuilt	1 rebuilt	D8340/17				*Kest* 5, Lahr-Dinglingen
D.III rebuilt	1 rebuilt	D8342/17		F.W. 93	10-Sep-18	*Kest* 4b, Freiburg
D.III rebuilt	1 rebuilt	D8343/17		F.W. 35	15-Aug-18	*Flugzeugmeisterei* Adlershof
D.III rebuilt	1 rebuilt	D8344/17		F.W. 25	10-Aug-18	*Kest* 5, Lahr-Dinglingen
D.III rebuilt	1 rebuilt	D8345/17		F.W. 73	6-Sep-18	*Kest* 8, Bitsch
D.III rebuilt	1 rebuilt	D8346/17		F.W. 66	2-Sep-18	*Kest* 4b, Freiburg
D.III rebuilt	1 rebuilt	D8347/17		F.W. 77	5-Sep-18	*Kest* 6, Bonn-Hangelar
D.III rebuilt	1 rebuilt	D8348/17		F.W. 102	11-Sep-18	*Kest* 4b, Freiburg
D.III rebuilt	1 rebuilt	D8349/17		F.W. 93	10-Sep-18	*Kest* 4b, Freiburg

	SSW D.III & D.IV Shipment Logs	
Engine Mfr. & Serial	**Remarks**	**References**
Sh.III(Alt)	Originally D.IIc(*kurz*) 7550/17. First flew 22-Oct-17	1
Sh.III	Originally D.IIc(*lang*) 7551/17. First flew 15-Nov-17. After crashing January 26, 1918 at the Adlershof Trials, was rebuilt as D.IV 7554/17	1
Sh.III	First flight 20-Dec-17. Crashed at January 1918 Adlershof Trials. Fuselage subsequently repaired at Blockwerk for experimental purposes. Written off July 1918 by order of *Flugzeugmeisterei*	1
Sh.III		15
Sh.III	Photographed at Siemensstadt, late Feb.–early Mar. 1918. Flat-topped rudder. May have been lost in British shelling of Balâtre airfield, April 12–13, 1918.	2, 15
Sh.III	Photo inscribed "*JG* II, Mesnil. *Oblt.* Krapfenbauer". Conflicts w/ consignee	3
Sh.III		
Sh.III		
Sh.III	Photo inscribed *Jasta* 19, *JG* II. Conflicts w/ consignee	3
Sh.III	Fuselage & fixed tail surfaces painted white. *Lt.* Goettsch, *Jasta* 19	15, 16
Sh.III		
Sh.III		
Sh.III		
Sh.III		
Sh.III		
Sh.III		
Sh.III		
Sh.III		
Sh.III		
Sh.III		
Sh.III		
Sh.III		
Sh.III		
Sh III		
	Vzfw. Fritz Beckhardt flew it with *Kest* 5, flew to Gossau, Switzerland and crash-landed 13 Nov 1918. Interned by Swiss. Does not appear on any SSW shipping records after 16 Mar 1918	
Sh.III *neu*		
Rh. 169	Possibly used to generate barograph @ Adlershof, Sept. 1918	
Rh. 171	*Offizieraspirant* Arnold Eger. Flown to Schaffhausen, Switzerland & crash-landed, 13 Nov. 1918. Interned by Swiss	
Rh. 228		
Sh.III *neu*		
Sh.III *neu*		
Sh.III *neu*		
Sh.III *neu*	Marked w/8-pointed star on fuselage, 9 on fin. Photographed in French hands after Armistice	

SSW D.III & D.IV Shipment Logs						
Type	**Series**	**Serial #**	**Order Date**	**Letter of Advice**	**Delivery Date**	**Consignee**
D.III rebuilt	1 rebuilt	D8350/17		F.W. 33	15-Aug-18	Mannheim (Rhemag?)
D.III rebuilt	1 rebuilt	D8351/17		F.W. 66	2-Sep-18	*Kest* 4b, Freiburg
D.III rebuilt	1 rebuilt	D8352/17		F.W. 78	6-Sep-18	*Kest* 8, Bitsch
D.III rebuilt	1 rebuilt	D8353/17		F.W. 25	10-Aug-18	*Kest* 5, Lahr-Dinglingen
D.III rebuilt	1 rebuilt	D8354/17		F.W. 55	24-Aug-18	*Kest* 8, Bitsch
D.III rebuilt	1 rebuilt	D8355/17		F.W. 19	3-Aug-18	*Kest* 5, Lahr-Dinglingen
D.III rebuilt	1 rebuilt	D8356/17		F.W. 73	4-Sep-18	*Kest* 5, Lahr-Dinglingen
D.III rebuilt	1 rebuilt	D8357/17		F.W. 66	2-Sep-18	*Kest* 4b, Freiburg
D.III rebuilt	1 rebuilt	D8358/17		F.W. 25	10-Aug-18	*Kest* 5, Lahr-Dinglingen
D.III rebuilt	1 rebuilt	D8359/17		F.W. 19	3-Aug-18	*Kest* 5, Lahr-Dinglingen
D.III	2	D1600/18	1-Mar-18	–	19-Apr-18	*JG* II
D.III	2	D1601/18	1-Mar-18	–	19-Apr-18	*JG* II
D.III	2	D1602/18	1-Mar-18	–	19-Apr-18	*JG* II
D.III	2	D1603/18	1-Mar-18	–	19-Apr-18	*JG* II
D.III	2	D1604/18	1-Mar-18	–	30-Apr-18	*JG* II
D.III	2	D1605/18	1-Mar-18	–	19-Apr-18	*JG* II
D.III	2	D1606/18	1-Mar-18	–	30-Apr-18	*JG* II
D.III	2	D1607/18	1-Mar-18	–	30-Apr-18	*JG* II
D.III	2	D1608/18	1-Mar-18	D391JSl	30-Apr-18	*JG* II
D.III	2	D1609/18	1-Mar-18	D389Sfl	29-May-18	*Z.A.K.* 3, Adlershof
D.III	2	D1610/18	1-Mar-18	D421Sfl	18-May-18	*JG* II
D.III	2	D1611/18	1-Mar-18	–	30-Apr-18	*JG* II
D.III	2	D1612/18	1-Mar-18	–	30-Apr-18	*JG* II
D.III	2	D1613/18	1-Mar-18	D421Sfl	18-May-18	*JG* II
D.III	2	D1614/18	1-Mar-18	–	13-May-18	*JG* II
D.III	2	D1615/18	1-Mar-18	–	13-May-18	*JG* II
D.III	2	D1616/18	1-Mar-18	–	13-May-18	*JG* II
D.III	2	D1617/18	1-Mar-18	–	13-May-18	*JG* II
D.III	2	D1618/18	1-Mar-18	–	13-May-18	*JG* II
D.III	2	D1619/18	1-Mar-18	–	13-May-18	*JG* II
D.III	2	D1620/18	1-Mar-18	–	22-May-18	–
D.III	2	D1621/18	1-Mar-18	D421Sfl	18-May-18	*JG* II
D.III	2	D1622/18	1-Mar-18	–	23-May-18	S.S.W.
D.III	2	D1623/18	1-Mar-18	–	23-May-18	–
D.III	2	D1624/18	1-Mar-18	D421Sfl	18-May-18	*JG* II
D.III	2	D1625/18	1-Mar-18	–	23-May-18	–
D.III	2	D1626/18	1-Mar-18	–	24-May-18	–
D.III	2	D1627/18	1-Mar-18	–	23-May-18	–

SSW D.III & D.IV Shipment Logs		
Engine Mfr. & Serial	**Remarks**	**References**
Less engine	Probably Udet's *LO!* Rhemag installed blueprinted engine, presented to Udet.	
Sh.III *neu*		
Rh. 207		
Rh. 187	War booty to Italy 10 July 20	
Sh.III *neu*		
Rh. 165		
Sh.III *neu*	*Gefr.* Bruno Lange flew to Switzerland 13 Nov.	3, 4, 5
Sh.III *neu*	Pilot: Weber. Black cowl & spinner	3
Rh. 158		
Rh. 168		
–		
–		
–		
–		
–		
–		
–		
–		
Sh.III		
Sh.IIIa	Test A/C. 8.600m upper, 8.220m lower span. Sketched 10-May & 5-June-1918	6
Sh.III		
–		
–		
Sh.III		
–		
–		
–		
–		
–		
–		
Sh. 9703	Possible subject S.S.W. sketch "D.III 6210/18", 15-May-18. 8.420m upper, 8.255m lower span (asymmetrical).	7
Sh.III		
–	Retained by S.S.W. & rebuilt w/ ailerons in upper wing only. Designated D.IIIa *"Eule"* (Owl)	1
–		
Sh.III		
–		
–		
–	Damaged during delivery flight	

Type	Series	Serial #	Order Date	Letter of Advice	Delivery Date	Consignee
D.III	2	D1628/18	1-Mar-18	–	27-May-18	–
D.III	2	D1629/18	1-Mar-18	–	27-May-18	–
D.III	2 rebuilt	D1600/18	–	F.W. 18	2-Aug-18	*Kest* 6, Bonn-Hangelar
D.III	2 rebuilt	D1601/18	–	F.W. 18	2-Aug-18	*Kest* 6, Bonn-Hangelar
D.III	2 rebuilt	D1602/18	–	F.W. 18	2-Aug-18	*Kest* 6, Bonn-Hangelar
D.III	2 rebuilt	D1603/18	–	D509Sfl	22-Jul-18	*JG* II über Saarbrücken
D.III	2 rebuilt	D1604/18	–	F.W. 89	7-Sep-18	*Kest* 4b, Freiburg
D.III	2 rebuilt	D1605/18	–	F.W. 55	24-Aug-18	*Kest* 8, Bitsch
D.III	2 rebuilt	D1606/18	–	F.W. 55	24-Aug-18	*Kest* 8, Bitsch
D.III	2 rebuilt	D1607/18	–	F.W. 25	10-Aug-18	*Kest* 5, Lahr-Dinglingen
D.III	2 rebuilt	D1608/18	–	–	–	–
D.III	2 rebuilt	D1609/18	–	–	–	–
D.III	2 rebuilt	D1610/18	–	–	–	–
D.III	2 rebuilt	D1611/18	–	F.W. 66	2-Sep-18	*Kest* 4b, Freiburg
D.III	2 rebuilt	D1612/18	–	F.W. 89	7-Sep-18	*Kest* 4b, Freiburg
D.III	2 rebuilt	D1613/18	–	–	–	–
D.III	2 rebuilt	D1614/18	–	D509Sfl	22-Jul-18	*JG* II über Saarbrücken
D.III	2 rebuilt	D1615/18	–	D509Sfl	22-Jul-18	*JG* II über Saarbrücken
D.III	2 rebuilt	D1616/18	–	D509Sfl	22-Jul-18	*JG* II über Saarbrücken
D.III	2 rebuilt	D1617/18	–	D509Sfl	22-Jul-18	*JG* II über Saarbrücken
D.III	2 rebuilt	D1618/18	–	F.W. 19	3-Aug-18	*Kest* 5, Lahr-Dinglingen
D.III	2 rebuilt	D1619/18	–	F.W. 18	2-Aug-18	*Kest* 6, Bonn-Hangelar
D.III	2 rebuilt	D1620/18	–	F.W. 26	10-Aug-18	*Kest* 5, Lahr-Dinglingen
D.III	2 rebuilt	D1621/18	–	–	–	–
D.III	2 rebuilt	D1622/18	–	D514Sfl	25-Jul-18	*Blockwerke*
D.III	2 rebuilt	D1623/18	–	F.W. 26	10-Aug-18	*Kest* 5, Lahr-Dinglingen
D.III	2 rebuilt	D1624/18	–	–	–	–
D.III	2 rebuilt	D1625/18	–	F.W. 74	4-Sep-18	*Kest* 5, Lahr-Dinglingen
D.III	2 rebuilt	D1626/18	–	F.W. 65	2-Sep-18	*Kest* 4b, Freiburg
D.III	2 rebuilt	D1627/18	–	F.W. 100	11-Sep-18	*Kest* 4b, Freiburg
D.III	2 rebuilt	D1628/18	–	F.W. 65	2-Sep-18	*Kest* 4b, Freiburg
D.III	2 rebuilt	D1629/18	–	F.W. 36	15-Aug-18	*Flugzeugmeisterei* Adlershof
D.III	3	D3007/18	23-Mar-18	F.W. 135	3-Oct-18	*Kest* 8, Bitsch
D.III	3	D3008/18	23-Mar-18	F.W. 80	6-Sep-18	*Kest* 8, Bitsch
D.III	3	D3009/18	23-Mar-18	F.W. 139	5-Oct-18	*Kest* 4a, Böblingen
D.III	3	D3010/18	23-Mar-18	F.W. 135	3-Oct-18	*Kest* 8, Bitsch

Engine Mfr. & Serial	Remarks	References
	**SSW D.III & D.IV Shipment Logs**	

Engine Mfr. & Serial	Remarks	References
–		
–		
Rh. 180		
Rh. 170		
Rh. 179		
Rh. 167		
Sh.III *neu*		
Rh. 190		
Sh.III *neu*		
Rh. 186		
–		
–		
–		
Sh.III *neu*	Pilot: Kessler. Dark spinner & cowling, serpent on fuselage, 6 on fin	3
Sh.III *neu*		
–		
Rh. 160		
Rh. 143		
Rh. 145		
Rh. 161		
Rh. 177	*Oblt.* Heinrich Dembrowsky flew to Schaffhausen, Switzerland and crash-landed, 13 Nov 1918. Interned by Swiss. #19188356 appears on wheel cover of 2nd or 3rd series D.III with "FR" monogram – probably incorrect	
Rh. 173		
Rh. 176	Photographed @ Siemensstadt. 5:4 *Balkankreuze* overpainted on fuselage & fin	3
–		
Sh.III	See entry for D1622/18 Series 2	
Rh. 162		
–		
Sh.III *neu*		
Sh.III *neu*	Pilot: *Vzfw.* Reimann. Dark & white diagonal stripes on tail	3
Sh.III		
Sh.III *neu*	Pilot: *Vzfw.* Paul Leim. Dark spinner, cowl & front fuselage panels. Black & white bands around fuselage	3
Sh.III *neu*		
Sh.III *neu*		
Rh. 181	Photographed @ Siemensstadt w/ square cowling cutaway & punched spinner	8
Sh.III *neu*		
Sh.III *neu*		

Type	Series	Serial #	Order Date	Letter of Advice	Delivery Date	Consignee
D.III	3	D3011/18	23-Mar-18	F.W. 139	11-Oct-18	*Kommandantur Flugplatz* Döberitz
D.III	3	D3012/18	23-Mar-18	F.W. 139	3-Oct-18	*Kest* 8, Bitsch
D.III	3	D3013/18	23-Mar-18	F.W. 212	5-Oct-18	*Kest* 4a, Böblingen
D.III	3	D3014/18	23-Mar-18	F.W. 212	5-Oct-18	*Kest* 4a, Böblingen
D.III	3	D3015/18	23-Mar-18	F.W. 212	8-Nov-18	*Kest* 2, Saarbrücken
D.III	3	D3016/18	23-Mar-18	F.W. 212	8-Nov-18	*Kest* 2, Saarbrücken
D.III	3	D3017/18	23-Mar-18	F.W. 212	8-Nov-18	*Kest* 2, Saarbrücken
D.III	3	D3018/18	23-Mar-18	F.W. 160	14-Oct-18	*Jastaschule* 1
D.III	3	D3019/18	23-Mar-18	F.W. 212	8-Nov-18	*Kest* 2, Saarbrücken
D.III	3	D3020/18	23-Mar-18	F.W. 212	8-Nov-18	*Kest* 2, Saarbrücken
D.III	3	D3021/18	23-Mar-18	F.W. 213	9-Nov-18	*Kest* 4a, Böblingen
D.III	3	D3022/18	23-Mar-18	F.W. 214	9-Nov-18	*Kest* 4a, Böblingen
D.III	3	D3023/18	23-Mar-18	L. FW. 4	14-May-19	*Fliegerhorst Brieg*
D.III	3	D3024/18	23-Mar-18	L. FW. 4	17-May-19	*Fliegerhorst Brieg*
D.III	3	D3025/18	23-Mar-18	F.W.99	9-Nov-18	*Kest* 8, Bitsch
D.III	3	D3026/18	23-Mar-18	F.W. 99	9-Nov-18	*Kest* 8, Bitsch
D.III	3	D3037/18	23-Mar-18	F.W. 140	4-Oct-18	*Kest* 6, Bonn
D.III	3	D3038/18	23-Mar-18	F.W. 160	14-Oct-18	*Jastaschule* 1
D.III	3	D3039/18	23-Mar-18	L. FW. 4	24-Jun-19	*Marinelandflieger Abt.*, Johannisthal
D.III	3	D3040/18	23-Mar-18	F.W. 75	4-Sep-18	*Kest* 5, Lahr-Dinglingen
D.III	3	D3041/18	23-Mar-18	F.W. 140	4-Oct-18	*Kest* 6, Bonn
D.III	3	D3042/18	23-Mar-18	F.W. 156	11-Oct-18	*Kommandantur Flugplatz* Döberitz
D.III	3	D3043/18	23-Mar-18	F.W. 165	16-Oct-18	*Kest* 4a, Böblingen
D.III	3	D3044/18	23-Mar-18	F.W. 165	16-Oct-18	*Kest* 4a, Böblingen
D.III	3	D3045/18	23-Mar-18	F.W. 140	4-Oct-18	*Kest* 6, Bonn
D.III	3	D3046/18	23-Mar-18	F.W. 139	5-Oct-18	*Kest* 4a, Böblingen
D.IV	Prototype	D7553/17	7-Oct-17	D509Sfl	15-Apr-18	*JG* II, 18.*Armee*
D.IV	Prototype	D7554/17	7-Oct-17	D505Sfl	23-Jul-18	*Autowerk*
D.IV	Prototype	D7555/17	7-Oct-17	F.W. 4	1-Aug-18	*Flugzeugmeisterei* Adlershof
D.IV	1	D3027/18	23-Mar-18	F.W. 52	23-Aug-18	*Jasta* 14, *Flugpark* 6
D.IV	1	D3028/18	23-Mar-18	F.W. 6	1-Aug-18	*Flugpark* 4, Ghent, *Marinekorps*

SSW D.III & D.IV Shipment Logs

Engine Mfr. & Serial	Remarks	References
Sh.III *neu*		
Sh.III *neu*		
Sh.III *neu*		
Sh.III *neu*		
Sh. 9588	Returned to factory after Armistice. Shipped to Döberitz 24 May 1919	
Sh. 9753	Returned to factory after Armistice. Shipped to Döberitz 22 or 31 May 1919	
Sh. 9702	Returned to factory after Armistice. Shipped to *Fliegerhorst Brieg* 14 May 1919	
Sh.III *neu*		
–	Returned to factory after Armistice. Shipped to *Fliegerhorst Brieg* 24 May 1919	
Sh. 9567	Returned to factory after Armistice. Shipped to *Fliegerhorst Brieg* 24 May 1919	
Sh. 9544	Returned to factory after Armistice. Shipped to *Fliegerhorst Brieg* 15 May 1919	
Sh. 9594	Returned to factory after Armistice. Shipped to *Fliegerhorst Brieg* 14 May 1919. War booty to Italy 25 June 1920	
Sh. 9830	As an experiment, equipped with Flettner ailerons	1
Sh. 9505	War booty to Italy 25 June 1920	
Rh. 182	Photographed in U.S. hands in Germany. White arrows on fuselage sides	
Rh. 216		
Sh.III *neu*		
Sh.III *neu*		
Sh. 9786		
Sh.III *neu*		
Sh.III *neu*		
Sh.III *neu*		
Sh.III *neu*	Acceptance sheet dated 26 September 1918. Fitted with Bamberg compass 30691, Bruhn tachometer #2375, Bosch starting magneto #35572, altimeter #15424, & machine guns #289 & 4363.	9
Sh.III *neu*	War booty to Italy 10 July 1920	
Sh.III *neu*		
Sh.III *neu*		
Rh. 159	Originally D.IIe 7553/17 w/ Dural wing spars. Crashed first flight 25 Jan.18. Re-built as D.IV, sent to *JG* II experimental detach. 15 May 18. Rebuilt @ SSW 6/18, sent to *JG* II, *Jasta* 12, Saarbrücken 22 July 18. Fuselage blue with white band	1, 13, 16
Sh.III		
Sh.III	First flight 18 June 1918. Prototype for D.IV Series	1
Rh. 181		
Rh. 166	Photographed at front with *Marine Landfeld Jasta*. One document says this plane reached 7,500 m in 39.6 min. at Mannheim, 26 July 1918, Lt. Müller piloting	3, 10

SSW D.III & D.IV Shipment Logs						
Type	Series	Serial #	Order Date	Letter of Advice	Delivery Date	Consignee
D.IV	1	D3029/18	23-Mar-18	F.W. 6	1-Aug-18	*Flugpark* 4, Ghent, *Marinekorps*
D.IV	1	D3030/18	23-Mar-18	F.W, 52	23-Aug-18	*Jasta* 14, *Flugpark* 6
D.IV	1	D3031/18	23-Mar-18	F.W. 52	23-Aug-18	*Jasta* 14, *Flugpark* 6
D.IV	1	D3032/18	23-Mar-18	F.W. 60	30-Aug-18	*Armee Flugpark* 4, *Marinekorps*
D.IV	1	D3033/18	23-Mar-18	F.W. 52	23-Aug-18	*Jasta* 14, *Flugpark* 6
D.IV	1	D3034/18	23-Mar-18	F.W. 6	1-Aug-18	*Armee Flugpark* 4, *Marinekorps*
D.IV	1	D3035/18	23-Mar-18	F.W. 59	30-Aug-18	*Jasta* 22, *Flugpark* 18
D.IV	1	D3036/18	23-Mar-18	F.W. 279	31-May-19	*Marine Landflieger Abt.*, Johannistal
D.IV	1	D3047/18	23-Mar-18	F.W. 52	23-Aug-18	*Jasta* 14, *Flugpark* 6
D.IV	1	D3048/18	23-Mar-18	F.W. 60	30-Aug-18	*Armee Flugpark* 4, *Marinekorps*
D.IV	1	D3049/18	23-Mar-18	–	–	–
D.IV	1	D3050/18	23-Mar-18	F.W. 52	23-Aug-18	*Jasta* 14, *Flugpark* 6
D.IV	1	D3051/18	23-Mar-18	F.W. 60	30-Aug-18	*Armee Flugpark* 4, *Marinekorps*
D.IV	1	D3052/18	23-Mar-18	F.W. 52	23-Aug-18	*Jasta* 14, *Flugpark* 6
D.IV	1	D3053/18	23-Mar-18	F.W. 60	30-Aug-18	*Armee Flugpark* 4, *Marinekorps*
D.IV	1	D3054/18	23-Mar-18	F.W. 87	16-Sep-18	*Flugpark* 4, Ghent, *Marinekorps*
D.IV	1	D3055/18	23-Mar-18	F.W. 59	30-Aug-18	*Jasta* 22, *Flugpark* 18
D.IV	1	D3056/18	23-Mar-18	F.W. 87	16-Sep-18	*Flugpark* 4, Ghent, *Marinekorps*
D.IV	2	D3060/18	7-May-18	F. W. 184	25-Oct-18	*JG* II
D.IV	2	D3061/18	7-May-18	F. W. 122	25-Sep-18	*Kest* 2, Saarbrücken
D.IV	2	D3062/18	7-May-18	F. W. 113	20-Sep-18	*Armee Flugpark* 4, *Marinekorps*
D.IV	2	D3063/18	7-May-18	F. W. 183	25-Oct-18	*JG* II
D.IV	2	D3064/18	7-May-18	F. W. 116	21-Sep-18	*Jasta* 22, *Flugpark* 18
D.IV	2	D3065/18	7-May-18	F. W. 116	21-Sep-18	*Jasta* 22, *Flugpark* 18
D.IV	2	D3066/18	7-May-18	F. W. 122	25-Sep-18	*Kest* 2, Saarbrücken
D.IV	2	D3067/18	7-May-18	F. W. 125	28-Sep-18	*Kest* 2, Saarbrücken
D.IV	2	D3068/18	7-May-18	F. W. 84	6-Sep-18	*Flugzeugmeisterei* Adlershof
D.IV	2	D3069/18	7-May-18	F. W. 144	7-Oct-18	*Jasta* 22, *Flugpark* 18
D.IV	2	D3070/18	7-May-18	F. W. 144	7-Oct-18	*Jasta* 22, *Flugpark* 18
D.IV	2	D3071/18	7-May-18	F. W. 125	28-Sep-18	*Kest* 2, Saarbrücken
D.IV	2	D3072/18	7-May-18	F. W. 113	20-Sep-18	*Armee Flugpark* 4, *Marinekorps*

SSW D.III & D.IV Shipment Logs		
Engine Mfr. & Serial	**Remarks**	**References**
Rh. 175	One document says *Flugpark* 2	
Sh.III		
Rh. 164		
Sh.III *neu*		
Rh. 188		
Rh. 163		
Sh.III *neu*		
Sh. 9832	Originally scheduled for shipment to Lagerhalle, Belgium, 28-Nov-18, on Lager F. W. 5. One document states 24-May-19 release per *Lt.* Rath (*notiz*)	
Rh. 185		
Sh.III *neu*	Assigned *Marine Landfeld Jasta*	
–	Wrecked on delivery flight – photos of wreck exist. Pilot: *Lt.* Speer.	3
Sh.III *neu*		
Sh.III *neu*		
Sh.III		
Sh.III *neu*		
Sh.III *neu*		
Sh.III *neu*		
Sh.III *neu*		
Rh. 310		
Sh.III *neu*		
Sh.III *neu*	*Flugpark* located at Hannover-Süd	
Rh. 285		
Rh. 192		
Rh. 220		
Sh.III *neu*		
Sh.III *neu*		
Sh.III *neu*	Probably the D.IV which was flown to 8,000m in 36 minutes at Aldershof on 16 Sept. 1918. Pilot: Rodschinka	11
Rh. 249		
Rh. 264		
Sh.III *neu*		
Sh.III *neu*	*Flugpark* located at Hannover-Süd	

Type	Series	Serial #	Order Date	Letter of Advice	Delivery Date	Consignee
D.IV	2	D3073/18	7-May-18	F. W. 105	16-Sep-18	*Kest* 2, Saarbrücken
D.IV	2	D3074/18	7-May-18	F. W. 79	8-Oct-18	*Jasta* 22, *Flugpark* 18
D.IV	2	D3075/18	7-May-18	F. W. 202	4-Nov-18	–
D.IV	2	D3076/18	7-May-18	F. W. 79	8-Oct-18	*Jasta* 22, *Flugpark* 18
D.IV	2	D3077/18	7-May-18	F. W. 167	17-Oct-18	*JG* II in field
D.IV	2	D3078/18	7-May-18	F. W. 148	9-Oct-18	*JG* II, *Jasta* 22, Park 18
D.IV	2	D3079/18	7-May-18	F. W. 164	16-Oct-18	*Flugplatz* Döberitz
D.IV	2	D3080/18	7-May-18	F. W. 163	16-Oct-18	*Flugzeugmeisterei* Adlershof, *ZAK* 5
D.IV	2	D3081/18	7-May-18	F. W. 148	9-Oct-18	*Jasta* 22, *Flugpark* 18
D.IV	2	D3082/18	7-May-18	F. W. 167	17-Oct-18	*JG* II in field
D.IV	2	D3083/18	7-May-18	F. W. 184	25-Oct-18	*JG* II
D.IV	2	D3084/18	7-May-18	F. W. 193	31-Oct-18	*Flugplatz* Döberitz
D.IV	2	D3085/18	7-May-18	F. W. 266	26-Mar-19	*Fl. Abt.* 431, Klein-Gandau (Breslau)
D.IV	2	D3086/18	7-May-18	F. W. 279	31-May-19	*Marine Landflieger Abt.*, Johannistal
D.IV	2	D3087/18	7-May-18	F. W. 265	22-Mar-19	*Fl. Abt.* 431, Klein-Gandau (Breslau)
D.IV	2	D3088/18	7-May-18	F. W. 288	24-Jun-19	*Marine Landflieger Abt.*, Johannistal
D.IV	2	D3089/18	7-May-18	F. W. 279	31-May-19	*Marine Landflieger Abt.*, Johannistal
D.IV	2	D3090/18	7-May-18	L. FW. 15	–	–
D.IV	2	D3091/18	7-May-18	L. FW. 15	–	–
D.IV	2	D3092/18	7-May-18	L. FW. 15	–	–
D.IV	2	D3093/18	7-May-18	L. FW. 15	–	–
D.IV	2	D3094/18	7-May-18	L. FW. 22	–	–
D.IV	2	D3095/18	7-May-18	L. FW. 22	–	–
D.IV	2	D3096/18	7-May-18	L. FW. 22	–	–
D.IV	2	D6150/18	8-Apr-18	F. W. 188	28-Oct-18	*Jasta* 14
D.IV	2	D6151/18	8-Apr-18	F. W. 200	2-Nov-18	*JG* II
D.IV	2	D6152/18	8-Apr-18	F. W. 182	25-Oct-18	*JG* II
D.IV	2	D6153/18	8-Apr-18	F. W. 157	12-Oct-18	*Kommandantur Flugplatz* Döberitz

SSW D.III & D.IV Shipment Logs		
Engine Mfr. & Serial	**Remarks**	**References**
Sh.III *neu*	One document says F. W. 110 dated 28-Sept-18	
Rh. 253		
Rh. 244	Wrecked on delivery flight. Pilot: *Lt*. Speer.	
Rh. 268		
Rh. 299	Found by French in hangar at Habay-la Neuve, Belgium, after armistice. Subject of *S.T.Aé.* report, November–December 1918	12
Rh. 239		
Rh. 249		
Rh. 280		
Rh. 236		
Rh. 278	Photographed @ Siemensstadt. Prop had small diameter, unpunched spinner marked 10237.	3
Rh. 234	Photographed with *Lt*. Lenz, *Jasta* 22	3
Sh.III *neu*		
Sh. 9592	Originally scheduled for shipment to Lagerhalle, Belgium, on 22-Nov-18, per letter of advice L. FW. 2	
Sh. 9506	Originally scheduled for shipment to Lagerhalle, Belgium, on 22-Nov-18, per letter of advice L. FW. 2	
Sh. 9855	Originally scheduled for shipment to Lagerhalle, Belgium, on 21-Dec-18, per letter of advice L. FW.11. War booty to Italy 25 June 1920	
Rh. 243	Originally scheduled for shipment to Lagerhalle, Belgium, on 22-Dec-18, per letter of advice L. FW. 13	
Sh. 9565	Originally scheduled for shipment to Lagerhalle, Belgium, on 22-Dec-18, per letter of advice L. FW. 13	
Rh. 247	Originally scheduled for shipment to Lagerhalle, Belgium, on 10-Jan-19, per letter of advice L. FW. 15. Finished in SSW warehouse	
Rh. 287	Originally scheduled for shipment to Lagerhalle, Belgium, on 10-Jan-19, per letter of advice L. FW. 15	
Sh. 9647	Originally scheduled for shipment to Lagerhalle, Belgium, on 10-Jan-19, per letter of advice L. FW. 15	
Sh. 9768	Originally scheduled for shipment to Lagerhalle, Belgium, on 10-Jan-19, per letter of advice L. FW. 15	
Sh. 9615	Originally scheduled for shipment to Lagerhalle, Belgium, on 31-Jan-19, per letter of advice L. FW. 22	
Sh. 9945	Originally scheduled for shipment to Lagerhalle, Belgium, on 31-Jan-19, per letter of advice L. FW. 22	
Sh. 9910	Originally scheduled for shipment to Lagerhalle, Belgium, on 31-Jan-19, per letter of advice L. FW. 22	
Sh.III *neu*		
Rh. 267		
Rh. 235		
Rh. 238		

SSW D.III & D.IV Shipment Logs						
Type	**Series**	**Serial #**	**Order Date**	**Letter of Advice**	**Delivery Date**	**Consignee**
D.IV	2	D6154/18	8-Apr-18	F. W. 188	28-Oct-18	*Jasta* 14
D.IV	2	D6155/18	8-Apr-18	F. W. 188	28-Oct-18	*Jasta* 14
D.IV	2	D6156/18	8-Apr-18	F. W. 203	5-Nov-18	*Jasta* 14
D.IV	2	D6157/18	8-Apr-18	L. FW. 9	24-Jun-19	*Kommandantur Flugplatz* Döberitz
D.IV	2	D6158/18	8-Apr-18	F. W. 194	31-Oct-18	*JG* II
D.IV	2	D6159/18	8-Apr-18	F. W. 194	31-Oct-18	*JG* II
D.IV	2	D6160/18	8-Apr-18	F. W. 190	31-Oct-18	*JG* II
D.IV	2	D6161/18	8-Apr-18	F. W. 200	2-Nov-18	*JG* II
D.IV	2	D6162/18	8-Apr-18	F. W. 208	8-Nov-18	*ZAK* 5, Adlershof
D.IV	3	D6163/18	29-Aug-18	F. W. 201	4-Nov-18	*JG* II
D.IV	3	D6164/18	29-Aug-18	F. W. 200	2-Nov-18	*JG* II
D.IV	3	D6165/18	29-Aug-18	F. W. 194	31-Oct-18	*JG* II
D.IV	3	D6166/18	29-Aug-18	F. W. 188	28-Oct-18	*Jasta* 14
D.IV	3	D6167/18	29-Aug-18	F. W. 194	31-Oct-18	*JG* II
D.IV	3	D6168/18	29-Aug-18	F. W. 201	4-Nov-18	*JG* II
D.IV	3	D6169/18	29-Aug-18	F. W. 252	12-May-19	*Kofl.* 17, Thorn
D.IV	3	D6170/18	29-Aug-18	F. W. 200	2-Nov-18	*JG* II
D.IV	3	D6171/18	29-Aug-18	F. W. 252	12-May-19	*Kofl.* 17, Thorn
D.IV	3	D6172/18	29-Aug-18	F. W. 203	5-Nov-18	*Jasta* 14
D.IV	3	D6173/18	29-Aug-18	F. W. 203	5-Nov-18	*Jasta* 14
D.IV	3	D6174/18	29-Aug-18	F. W. 203	5-Nov-18	*Jasta* 14
D.IV	3	D6175/18	29-Aug-18	F. W. 252	12-May-19	*Kofl.* 17, Thorn
D.IV	3	D6176/18	29-Aug-18	F. W. 203	5-Nov-18	*Jasta* 14
D.IV	3	D6177/18	29-Aug-18	F. W. 252	12-May-19	*Kofl.* 17, Thorn. Later in Lithuanian possession
D.IV	3	D6178/18	29-Aug-18	F. W. 265	22-Mar-19	*Fl. Abt.* 431, Klein-Gandau (Breslau)
D.IV	3	D6179/18	29-Aug-18	F. W. 203	5-Nov-18	*Jasta* 14
D.IV	3	D6180/18	29-Aug-18	F. W. 252	12-May-19	*Kofl.* 17, Thorn
D.IV	3	D6181/18	29-Aug-18	F. W. 286	24-Jun-19	*Kommandantur Flugplatz* Döberitz
D.IV	3	D6182/18	29-Aug-18	F. W. 275	27-Jun-19	*Flz.* Aldershof, *Abt.* XII (*Versuche*)
D.IV	3	D6183/18	29-Aug-18	F. W. 252	12-May-19	*Kofl.* 17, Thorn

Engine Mfr. & Serial	Remarks	References
Sh.III *neu*		
Sh.III *neu*		
Sh. 9877	Some documents list ship date as 6-Nov-18	
Rh. 242	Originally scheduled for shipment to Lagerhalle, Belgium, on 21-Dec-18, per letter of advice L. FW. 9	
Sh.III *neu*		
Sh.III *neu*		
Sh.III *neu*		
Rh. 221B		
Rh. 341	Over-dimensioned Rhemag Sh.III	
Rh. 276		
Rh. 291		
Sh.III *neu*		
Sh.III *neu*		
Sh.III *neu*		
Rh. 281		
Rh. 294	Originally scheduled for shipment to Lagerhalle, Belgium, on 22-Nov-18, per letter of advice L. FW. 1. One document says shipped 12-Feb-19	
Rh. 309		
Sh. 9880	Originally scheduled for shipment to Lagerhalle, Belgium, on 22-Nov-18, per letter of advice L. FW. 1. One document says shipped 12-Feb-19	
Sh. 9829	Some documents list ship date as 6-Nov-18	
Sh. 9526	Photographed with SSW test pilot Rodschinka. Some documents list ship date as 6-Nov-18	3
Sh. 9688	Some documents list ship date as 6-Nov-18	
Sh. 9661	Originally scheduled for shipment to Lagerhalle, Belgium, on 22-Nov-18, per letter of advice L. FW. 1. One document says shipped 12-Feb-19	
Sh. 9515	Some documents list ship date as 6-Nov-18	
Sh. 9516	Originally scheduled for shipment to Lagerhalle, Belgium, on 22-Nov-18, per letter of advice L. FW. 1. One document says shipped 12-Feb-19	14
Sh. 9566	Originally scheduled for shipment to Lagerhalle, Belgium, on 21-Dec-18, per letter of advice L. FW. 9. Crashed 24-July-19, Hundsfeld bei Breslau, *Uffz.* Hassenmitter	3
Sh. 9739	Some documents list ship date as 6-Nov-18	
Sh. 9523	Originally scheduled for shipment to Lagerhalle, Belgium, on 22-Nov-18, per letter of advice L. FW. 1. One document says shipped 12-Feb-19	
Sh. 9878	Originally scheduled for shipment to Lagerhalle, Belgium, on 21-Dec-18, per letter of advice L. FW. 9. One document says shipped 19-Mar-19 to *Fl. Abt.* 431.	
Sh. 9890	Originally scheduled for shipment to Lagerhalle, Belgium, on 21-Dec-18, per letter of advice L. FW. 9	
Sh. 9755	Originally scheduled for shipment to Lagerhalle, Belgium, on 22-Nov-18, per letter of advice L. FW. 1	

SSW D.III & D.IV Shipment Logs						
Type	**Series**	**Serial #**	**Order Date**	**Letter of Advice**	**Delivery Date**	**Consignee**
D.IV	3	D6184/18	29-Aug-18	F. W. 278	31-May-19	*Kommandantur Flugplatz* Döberitz
D.IV	3	D9000/18	29-Aug-18	–	–	SSW Warehouse
D.IV	3	D9001/18	29-Aug-18	F. W. 274	23-May-19	*Fliegerhorst* Liegnitz
D.IV	3	D9002/18	29-Aug-18	F. W. 274	23-May-19	*Fliegerhorst* Liegnitz
D.IV	3	D9003/18	29-Aug-18	F. W. 289	24-Jun-19	*Kommandantur Flugplatz* Döberitz
D.IV	3	D9004/18	29-Aug-18	F. W. 277	28-May-19	*Fliegerhorst* Liegnitz
D.IV	3	D9005/18	29-Aug-18	F. W. 286	24-Jun-19	*Kommandantur Flugplatz* Döberitz
D.IV	3	D9006/18	29-Aug-18	F. W. 280	5-Jun-19	*Fliegerhorst* Liegnitz
D.IV	3	D9007/18	29-Aug-18	F. W. 277	28-May-19	*Fliegerhorst* Liegnitz
D.IV	3	D9008/18	29-Aug-18	F. W. 277	28-May-19	*Fliegerhorst* Liegnitz
D.IV	3	D9009/18	29-Aug-18	F. W. 286	24-Jun-19	*Kommandantur Flugplatz* Döberitz
D.IV	3	D9010/18	29-Aug-18	F. W. 292	31-Jul-19	*Fliegerhorst* Liegnitz
D.IV	3	D9011/18	29-Aug-18	–	16-Jan-19	SSW Warehouse
D.IV	3	D9012/18	29-Aug-18	F. W. 274	23-May-19	*Fliegerhorst* Liegnitz
D.IV	3	D9013/18	29-Aug-18	F. W. 274	23-May-19	*Fliegerhorst* Liegnitz
D.IV	3	D9014/18	29-Aug-18	F. W. 286	24-Jun-19	*Kommandantur Flugplatz* Döberitz
D.IV	3	D9015/18	29-Aug-18	F. W. 286	24-Jun-19	*Kommandantur Flugplatz* Döberitz
D.IV	3	D9017/18	29-Aug-18	F. W. 286	24-Jun-19	*Kommandantur Flugplatz* Döberitz
D.IV	3	D9018/18	29-Aug-18	F. W. 286	24-Jun-19	*Kommandantur Flugplatz* Döberitz
D.IV	3	D9019/18	29-Aug-18	F. W. 286	24-Jun-19	*Kommandantur Flugplatz* Döberitz
D.IV	3	D9020/18	29-Aug-18	–	31-Jan-19	SSW Warehouse
D.IV	3	D9022/18	29-Aug-18	F. W. 280	5-Jun-19	*Fliegerhorst* Liegnitz

	SSW D.III & D.IV Shipment Logs	
Engine Mfr. & Serial	**Remarks**	**References**
Sh. 9722	Originally scheduled for shipment to Lagerhalle, Belgium, on 22-Nov-18, per letter of advice L. FW. 1	
Sh. 9874	Originally scheduled for shipment to Lagerhalle, Belgium, on 21-Dec-18, per letter of advice L. FW. 10	
Sh. 9692	Originally scheduled for shipment to Lagerhalle, Belgium, on 21-Dec-18, per letter of advice L. FW. 10	
Sh. 9729	Originally scheduled for shipment to Lagerhalle, Belgium, on 21-Dec-18, per letter of advice L. FW. 10	
Rh. 222	Originally scheduled for shipment to Lagerhalle, Belgium, on 10-Jan-19, per letter of advice L. FW. 16	
Rh. 231	Originally scheduled for shipment to Lagerhalle, Belgium, on 22-Dec-18, per letter of advice L. FW. 12	
Sh. 9586	Originally scheduled for shipment to Lagerhalle, Belgium, on 22-Dec-18, per letter of advice L. FW. 12	
Rh. 223	Originally scheduled for shipment to Lagerhalle, Belgium, on 22-Dec-18, per letter of advice L. FW. 12	
Rh. 225	Originally scheduled for shipment to Lagerhalle, Belgium, on 22-Dec-18, per letter of advice L. FW. 12	
Sh. 9618	Originally scheduled for shipment to Lagerhalle, Belgium, on 21-Dec-18, per letter of advice L. FW. 10	
Rh. 237	Originally scheduled for shipment to Lagerhalle, Belgium, on 22-Dec-18, per letter of advice L. FW. 12	
Sh. 9659	Originally scheduled for shipment to Lagerhalle, Belgium, on 16-Jan-19, per letter of advice L. FW. 17	
Sh. 9887	Originally scheduled for shipment to Lagerhalle, Belgium, on 16-Jan-19, per letter of advice L. FW. 17	
Sh. 9684	Originally scheduled for shipment to Lagerhalle, Belgium, on 16-Jan-19, per letter of advice L. FW. 17	
Rh. 213	Originally scheduled for shipment to Lagerhalle, Belgium, on 16-Jan-19, per letter of advice L. FW. 17	
Sh. 9984	Originally scheduled for shipment to Lagerhalle, Belgium, on 31-Jan-19, per letter of advice L. FW. 20	
Rh. 256		
Rh. 262	Originally scheduled for shipment to Lagerhalle, Belgium, on 31-Jan-19, per letter of advice L. FW. 20	
Rh. 275	Originally scheduled for shipment to Lagerhalle, Belgium, on 31-Jan-19, per letter of advice L. FW. 20	
Sh. 9573	Originally scheduled for shipment to Lagerhalle, Belgium, on 31-Jan-19, per letter of advice L. FW. 20	
Rh. 266	Originally scheduled for shipment to Lagerhalle, Belgium, on 31-Jan-19, per letter of advice L. FW. 20	
Rh. 298	Originally scheduled for shipment to Lagerhalle, Belgium, on 31-Jan-19, per letter of advice L. FW. 20	

SSW D.III & D.IV Shipment Logs						
Type	**Series**	**Serial #**	**Order Date**	**Letter of Advice**	**Delivery Date**	**Consignee**
D.IV	3	D9029/18	29-Aug-18	F. W. 277	28-May-19	*Fliegerhorst* Liegnitz
D.IV	3	D11500/18	29-Aug-18	–	14-Dec-18	SSW Warehouse
D.IV	3	D11501/18	29-Aug-18	–	14-Dec-18	SSW Warehouse
D.IV	3	D11502/18	29-Aug-18	F. W. 292	31-Jul-19	*Fliegerhorst* Liegnitz

References: SSW factory production & shipping documents, including:

"Flugzeugwerk"
Lieferungen von SSW-Flugzeugen
Zusammenstellung von der SSW gelieferten D III und D IV Flugzeuge
Aufstellung über die bis zum 30.8.18 mit weisen Motoren (Rhemag oder SH 3) zur Ablieferung gekommenen D Flug-zeuge
Aufstellung zu unserer Rechnung FW0241 vom 30.6.19 über die erfolgten Lieferungen
Aufstellung zu unserer Rechnung FW0298 vom 31.7.19 über die erfolgten Lieferungen
Aufstellung zu unserer Rechnung FW0301 vom 31.7.19 über die erfolgten Lieferungen der Flugzeuge D IV 9000–9015, 9017–9020, 9022 und 9029
(BZ 11115) Aufstellung über die jenigen Reserve – Propeller bezw. Reserve – Propellerhauben, die den fertiggestellten 37 Flugzeugen: D IV 3060-3096 beigegeben wurden. Auftragsschreiben der Idflieg vom 7.5.18 B-No. 168/4.18 Flz A1.
(BZ 11163) Aufstellung über die den fertigsstelten Flugzeugen D IV 6150–6184 beigegebenen Reserve - Propeller und Reserve - Propellerhauben, Auftragsschr. der Idflieg vom 29.9.18 B-No. 683/7.18 Flz A1.

Note: In the shipment logs above, Sh.III *neu* is an alternative designation for Sh.IIIa, the modified Sh.III engine.

Below: SSW D.III under postwar evaluation at the French test center at Villacoublay.

Engine Mfr. & Serial	SSW D.III & D.IV Shipment Logs	References
	Remarks	
Sh. 9610	Originally scheduled for shipment to Lagerhalle, Belgium, on 31-Jan-19, per letter of advice L. FW. 20	
Sh. 9889	Originally scheduled for shipment to Lagerhalle, Belgium, on 13 or 14-Dec-18, per letter of advice L. FW. 7	
Sh. 9758	Originally scheduled for shipment to Lagerhalle, Belgium, on 13 or 14-Dec-18, per letter of advice L. FW. 7	
Sh. 9908	Originally scheduled for shipment to Lagerhalle, Belgium, on 13 or 14-Dec-18, per letter of advice L. FW. 7. One document says Liegnitz was final destination. Flown in by *Lt.* Rath 7-July-19. War booty to Italy 25 June 1920	

References Listed in Table:

1. Untitled typewritten list of SSW aircraft.
2. Photos of D 8341/17. RLB & PMG collections.
3. Photos. PMG collection.
4. Photo, Profile 86, p.12.
5. Photo, *Scale Models* magazine, July 1981, p.352.
6. Unterabteilung für Flugzeug-Festigkeitsprufung sketches dated 10 V 18 & 5 VI 18.
7. SSW sketch "D III 6120/18" dated 15 May 1918.
8. Photo, Profile 86, p.7.
9. Acceptance sheet for D. III 3043/18.
10. Photo, Profile 86, p.9.
11. *Flight*, March 13, 1919, p.339.
12. S.T.Aé. Rapport Mensuel, Nov–Dec, 1918.
13. *Technische Berichte Band III/Heft 5*: "Neuere Bestrebungen und Erfahrungen in Flugmotorenbau", Schwager, 1918, p.144.
14. "Pirmieji lietuvos karo avaicijos léktuval", Gytis Ramoska, Plieno Sparnai No. 1, Kaunas, 1992.
15. Imrie, Alex, *The Fokker Triplane*, Arms & Armour Press, London, 1992.
16. Grosz, P. M., *SSW D.III–D.IV*, Windsock Datafile 29, Albatros Publications, Berkhamsted, 1991.

Above: SSW D.III 8340/17 of *Kest* 5, the first production D.III fighter, is shown after *Vzfw.* Fritz Beckhardt's crash landing at Gossau, near Rapperswill, Switzerland, Nov. 13, 1918.

Above Right: SSW D.III 1618/18 after *Oblt.* Heinrich Dembrowsky crash-landed this aircraft in Schaffhausen, Switzerland, November 13, 1918.

Right: SSW D.III 8344/17 after being completely repainted in Swiss markings. This aircraft was retired from Swiss service in 1922 after lack of spares became a problem. It was scrapped and its engine and prop (cut down) were displayed at the Swiss Verkehrshaus museum.

SSW D.VI

Above: The SSW D.VI was not completed until after the armistice; by then the 'E' category for monoplane fighters was no longer used. Powered by the 205 hp Sh.IIIa high-altitude counter-rotary engine, it had a top speed of 220 km/h (137 mph) coupled with excellent maneuverability and climb and a ceiling of 8,000m. The under-fuselage fuel tank could be jettisoned in case of fire. The SSW D.VI was the natural production successor to the SSW D.IV.

As early as June 1917, Siemens-Schukert presented its project drawings for a parasol monoplane fighter, the SSW D.Ic, to a conference of interested *Idflieg* officials. The parasol project was developed by *Dipl.-Ing.* Glöckner, who was the deputy to the Chief Engineer of the SSW aircraft section, *Dipl.-Ing.* Harald Wolff. But because the triplane vogue was enjoying its peak popularity in *Idflieg* circles at this time, the parasol design was rejected.

It was not until April 1918 that SSW succeeded in obtaining a contract from *Idflieg* for construction of three experimental parasol fighters. The drawings were presented to *Idflieg* in June, at which time the company was allowed, at its request, to substitute wooden wings for the duraluminum wings originally planned. Several months previously SSW had experienced difficulty with the dural spars used in the thin profile wings of an experimental SSW D.III fighter. Now work could proceed on the detail drawings and a full-size mock-up of the parasol fighter was built, followed by construction of the three prototypes.

On August 12, 1918 *Idflieg* assigned the designation E.IV to the new fighter, but this designation was changed to D.VI the next month when *Idflieg* abandoned the E-type designation as

obsolete. It was intended to replace the SSW D.IV with D.VI series production provided the D.VI had the required performance. However, work on the prototypes proceeded slowly, and at the time of the Armistice only one prototype was almost complete. Of the other two prototypes, *Idflieg* decreed that only the one nearest completion be finished. The two D.VI parasol fighters were given serial numbers 3054/17 and 3055/17, numbers originally assigned to the earlier SSW Dr.II triplane project that had been cancelled.

At the end of January 1919 the first D.VI, 3054/17, was shipped to Berlin-Staaken, assembled there, and made its first test flight on February 3, 1919. On March 12, 1919 *Lt.* Hans-Joachim Rath of the Government Inspection office at SSW took off on the acceptance flight. In reaching the altitude of 7,200 meters, the D.VI showed a climb of 6,000 meters in 16 minutes and 7,000 meters in 22 minutes. Total flying weight of 3054/17 was 705.5 kg.

As was the case with the SSW D.III, D.IV, and D.V, the left wing of the D.VI was about 10 cm longer than the right in order to cancel the torque effect. However, in the case of the D.VI, the left wing was originally too long and had to be shortened to correct difficulties experienced during right turns.

The acceptance flight of the second D.VI, 3055/17,

Above: The SSW D.VI parasol monoplane undergoing static load testing at the factory. Bags full of sand are being used to distribute the load along the inverted wing to simulate aerodynamic loads in flight.

also piloted by *Lt*. Rath, took place on April 2, 1919 at Staaken. The measured climb was 6,000 meters in 20.4 minutes. The take-off run was measured at about 30 meters and landing run at about 80 meters. The measured speeds in level flight were:

Altitude:	2,000 m	220 km/h
	3,000 m	210 km/h
	4,000 m	190 km/h
	5,000 m	185 km/h

On 30 April 1919, *Lt*. Mohnike took off in 3055/17 to ferry the aircraft to the *Flugzeugmeisterei* (Aircraft Testing Section at *Idflieg*) at Adlershof. During the flight the aircraft caught fire; the pilot, a well-known and experienced fighter pilot, dove to the ground and landed safely, although the aircraft was totally destroyed. *Lt*. Mohnike had forgotten to release the jettisonable fuel tank (German Patent 342 084). Examination of the wreckage showed the fuel lines were all in order, but the solder joint on the emergency fuel tank's strainer had broken and could have allowed fuel to leak over the carburetor into the aircraft.

In May 1919 the other D.VI, 3054/17, was also involved in an accident. After many flights the aircraft was flown by an *Idflieg* pilot to Döberitz. During landing the left wing broke off, fracturing at the strut attachment points. After a thorough examination the accident commission came to the conclusion that the wing had been damaged prior to the flight by carelessness or sabotage. At this time, sabotage was a strong possibility given the turbulent political conditions in post-war Germany.

The damaged wing was repaired with the permission of the *Flugzeugmeisterei*. The airplane was to be sent to Adlershof for static load tests in July 1919 but these never occurred. At this time there were organizations in Germany who wanted to save valuable aircraft and not surrender them to the Allied Armistice Commission, so for a time the last D.VI was hidden in Berlin-Siemensstadt. However, the chauffeur of Herr von Siemens sold information about the hiding place to the Armistice Commission, whereupon the D.VI was found and destroyed.

This Page: The SSW D.VI parasol monoplane demonstrated excellent speed, climb, and ceiling and there is little doubt it would have replaced the SSW D.IV biplane in production in early 1919 had the war continued. Continued engine experience to verify reliability may have enabled it to serve as a general-purpose fighter as well as an interceptor, perhaps filling the role intended for the troubled Rumpler D.I.

Above: This close-up of the SSW D.VI shows the exhaust outlet and the jettisonable belly fuel tank to advantage. The D.VI shows careful attention to detail, especially the complex wing design. The wing was thickest where the bracing struts attached, giving it greatest strength at the area of greatest bending moment.

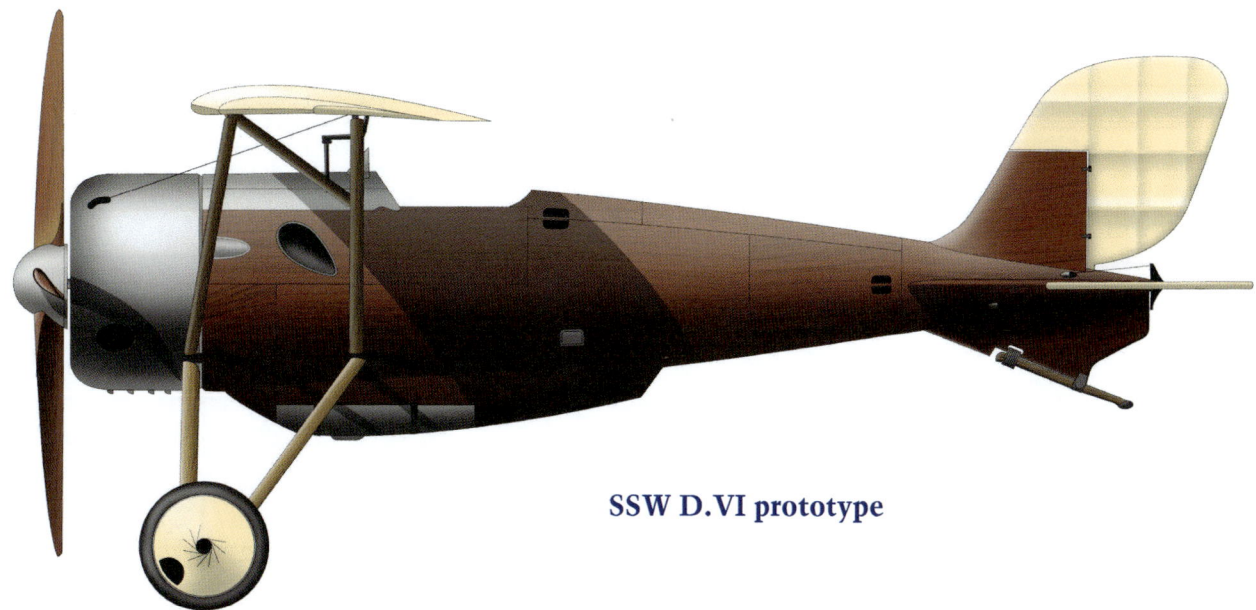

SSW D.VI prototype

Siemens-Schuckert Fighter Aircraft Summary									
		Upper Wing			Lower Wing			Total Wing	
Model	Serials	Span, m	Chord, m	Area, m²	Span, m	Chord, m	Area, m²	Area, m²	Length, m
E.I	550, 551, 553–558/15								
E.I	559–569/15								
E.II	552/25								
E.III	620–625/15								
D.I		7.50	1.3	9.8	6.3	0.8	5.04	14.40	
D.Ia		8.00	1.4	11.3	6.4	0.8	5.12	16.00	
D.Ib #1		8.00	1.5	11.6	6.4	0.8	5.12	16.20	
D.Ib #2		10.00	1.5	14.5	6.4	0.8	5.12	19.20	
D.II		7.80	1.6		7.4	0.9		18.70	
D.II		8.00	1.6	12.8	6.4	0.9	5.76	18.20	
D.IIa larger wing		15.00	1.0	15.0	6.4	0.8	5.12	19.70	
D.IIb		9.64	1.6		7.0	0.8		19.40	
D.IIb Alt.		8.00	1.5	12.0	6.4	0.8	5.12	16.70	
D.IIc kurz (later D.III)	7550/17	8.50			<8.5			19.40	6.00
D.IIc lang (later D.III)	7551/17	9.00						18.02	6.00
D.IIe (later D.III)	7553/17	8.7 (8.2)						15.40	6.00
D.III	7550/17	8.33	1.46		7.50	1.0		19.24	
D.III	7552/17	8.70						21.60	6.00
D.III	7552/17	9.05	1.20		8.70	1.0		18.33	
D.III	7552/17	8.40	1.47		8.28	1.01		18.73	5.75
D.IIIa	7552/17							19.40	6.00
D.III	7553/17								
D.III		8.00	1.00		8.00	1.0		15.20	
D.III	8340–8359/17	8.30	1.46		8.28	1.0		18.84	5.65
D.III rebuilt	8340, 8342–8349, 8351–8359/17	8.35	1.46		8.13	1.0			5.89
D.III	8350/17								
D.III	1600-1619, 1621, 1624	8.50	1.46	11.8	8.26	1.0	7.12	18.92	
D.III	1609/18	8.60	1.47	12.1	8.22	1.0	7.12	19.22	

Siemens-Schuckert Fighter Aircraft Summary

Weight, kg	Engine(s), hp	Remarks
	Sh.0@90	
	Sh.I@110	
	Argus As.II @120	
	Oberursel U.I@100	
660	Sh.I@110	
640	Sh.I@110	
600	Sh.Ia@125	High compression engine
620	Sh.Ia@125	High compression engine
/750	Sh.III@205	
745	Sh.III@205	
720	Sh.III@205	
/700	Sh.III@205	
700	Sh.III@205	
500/750	Sh.III@205	Steel tube V-struts, First flight 22 Oct. 1917. Climb to 5,000 m in 15 min., 6,000 m in 26.2 min.
500/750	Sh.III@205	Wooden V-struts, climb 5,000 m in 17.5 min. Crashed at Adlershof Jan. 1918. After crash, rebuilt as D.IV 7554/17.
500/	Sh.III@205	Dural spars, I-struts, no bracing. Crashed on 1st flight, 25 Jan 1918. Rebuilt as D.IV 7553/17
/715	Sh.III@205	First flight 22 Jan. 1918
500/750	Sh.III@205	First flight 20 Dec. 1917. Climb to 5,000 m in 12.5 min., 6,000 m in 21.5 min. Crashed during Jan. 1918 Adlershof Fighter Trials
708	Sh.III@205	"With wider span"
	Sh.III@205	First fighter Trials, Adlershof, January 1918. Measured for *Baubeschreibung*
	Sh.III@205	Abandoned July 1918 on order of *Flugzeugmeisterei*
/700	Sh.III@205	Climb to 6,000 m in 17 min. Sent to *JG* II for evaluation April 1918. Returned to factory, rebuilt & returned to *JG* II July 1918
	Sh.III@205	Test prototype "for higher speed"
515 (534)/725	Sh.III@205	First production series. Full cowling, inset aileron balances. Climb to 5,000 m in 13 min., 6,000 m in 17 min.
	Sh.IIIa@205	Rebuilt first series with cutaway cowling, elephant-ear ailerons & redesigned engine (Sh.IIIa)
	Sh.IIIa(Rh) @205	Shipped less engine to Rhemag. After engine installation, sent to *Lt.* Ernst Udet, *Jasta* 4
	Sh.IIIa@205	Second production series before cowling, aileron & engine changes
/749	Sh.IIIa@205	Shipped to *Z.A.K.* 3, Adlershof, for evaluation

Siemens-Schuckert Fighter Aircraft Summary										
		Upper Wing			Middle Wing			Lower Wing		
Model	Serials	Span, m	Chord, m	Area, m²	Span, m	Chord, m	Area, m²	Span, m	Chord, m	Area, m²
D.III	1620/18	8.458	1.46					8.255	1.0	
D.III	1620/18	8.260	1460.00					7.345	1.0	
D.III	1629/18									
D.IIIa	1622/18									
D.III rebuilt	1600–07, 1611, 1612, 1614–20, 1623, 1625–29									
D.III	3007–3026/18	8.43						8.43	1.0	
D.III	3008/18									
D.III	3023/18									
D.IV prototype		6.00	1.00					6.00	1.00	
D.IV prototype		8.00	1.00					8.00	1.00	
D.IV prototype		9.00	1.00					9.00	1.00	
D.IV prototype		12.00	1.00					12.00	1.00	
D.IV prototype		9.00	0.75					9.00	0.75	
D.IV prototype		9.00	1.00					9.00	1.00	
D.IV prototype		9.00	1.50					9.00	1.50	
D.IV	7555/17	8.20								
D.IV	7554/17	>7.4								
D.IV	3027–36/18, 3047–56, 3060–96, 6150–84, 9000–29, 11500–02	8.35	1.00					8.35	1.0	
D.IVa	7554/17	7.40						7.40		
D.V	7556–7558/17	8.86								
D.VI	3054–3055/18	9.37								
Dr.I #1		8.60	1.00	8.60	7.80	0.72	5.62	7.80	0.72	5.62
Dr.Ia		8.60	1.10	9.46	7.80	0.80	6.24	7.80	0.80	6.24
Dr.II		10.00	1.20	12.00	9.00	0.85	7.65	9.00	0.85	7.65
DDr.I		11.00	1.20	13.20	10.00	0.85	8.50	10.00	0.85	8.50
DDr.II										

Siemens-Schuckert Fighter Aircraft Summary

Total Wing Area, m²	Length, m	Weight, kg	Engine(s), hp	Remarks
			Sh.III@205	Apparently retained by SSW for testing. Asymmetrical lower wing; port span 4.125m, starboard span 4.140m
				Asymmetrical lower wing; port span 3.66 m, starboard 3.685 m
			Sh.III@205	High speed propeller. Apparently retained by SSW for testing.
			Sh.III@205	Nicknamed "Eule" (Owl). Ailerons on upper wing only. Otherwise standard 2nd Series D.III. Retained by SSW
18.84	5.70		Sh.IIIa@205	
			Sh.IIIa@205	Fitted with experimental half cowling
			Sh.IIIa@205	Experimental aircraft with Flettner ailerons. Otherwise, standard 3rd Series D.III
12.00		/730	Sh.III@205	Identified as 190 PS Doppeldecker. Probably part of a design study for the D.IV
15.20		/720	Sh.III@205	Probably part of a design study for the D.IV
18.00		/760	Sh.III@205	Identified as 190 PS Doppeldecker. Probably part of a design study for the D.IV
24.00		/790	Sh.III@205	Identified as 190 PS Doppeldecker. Probably part of a design study for the D.IV
13.50		/736	Sh.III@205	Identified as 190 PS Doppeldecker. Probably part of a design study for the D.IV
18.00		/760	Sh.III@205	Identified as 190 PS Doppeldecker. Probably part of a design study for the D.IV
27.00		/805	Sh.III@205	Identified as 190 PS Doppeldecker. Probably part of a design study for the D.IV
15.40	6.00	500/715	Sh.III@205	Prototype for D.IV series. First flight 18 June 1918. Shipped to Flugzeugmeisterei for destruction tests
		/695	Sh.IIIa@205	Formerly D.IIc lang 7551/17. After second crash, rebuilt as D.IVa
15.12	5.70	540 (542)/735	Sh.IIIa@205	First production series. Asymmetrical wings: port span 4.18 m, starboard span 4.17 m
14.00		/695	Sh.IIIa@205	Rebuilt from D.IIc lang 7551/17. Ailerons in upper wing only. Climb to 6,000m in 18 min. At Second Adlershof Fighter Trials
15.10	5.70	514/735	Sh.IIIa@205	Twin-bay wings with aluminum spars. 7558 first flight 14 June '18
12.46	6.50		Sh.IIIa@205	Parasol monoplane. First flight Februry 3, 1919
22.74		600	Sh.Ia@125	High compression engine, area of 4th wing 2.9 m²
24.84		630	LeRhone@140	Le Rhone engine, area of 4th wing 2.9 m²
(26.1) 27.3		495/700	Sh.III@205	Fuselage salvaged for D.IIb
(30) 33.7	5.80	680/910	2xSh.Ia@125	High compression engine, area of 4th wing 4.2 m². Crashed 1st flt
		869/1130	2xSh.III@205	Projected triplane development of DDr.I. Never built

SSW Guided Missiles

Guided missiles are not the kind of 'aircraft' one normally associates with WWI, but SSW built and tested them and they are included here for their interest. In October 1914 Dr. Wilhelm von Siemens proposed remotely-controlled glide bombs, an air to surface missile. Work began that month and by January 1915 small models were under test.

Scale model glide-bombs were launched from towers, balloons, and a track built on the roof of the Siemens airship shed at Biesdorf in addition to launches from airships and airplanes. Initially the gliders' servo-controls were powered by batteries, but these heavy cells were soon replaced by propeller-driven dynamos that generated electricity. Guidance commands were transmitted from airships to the glide bombs by thin wires that unraveled from a spool after the glider was launched. After guidance commands the rudder returned to neutral but the elevator remained in the last position commanded. Many flight tests with gliders from 92 to 265 pounds

Left: After being modified to eliminate its protruding belly, the SSW Bulldog, powered by one of the first 110 hp Siemens-Halske Sh.I counter-rotary engines, was used to launch scale gliders to develop shapes for the SSW guided missile program.

Right: Another view of a test glider mounted under the modified, rotary-powered SSW Bulldog before a test launch.

weight were made from autumn 1915 to spring 1917.

Ships are difficult to hit with conventional bombs due to their ability to maneuver quickly after bombs are released, so they were a natural target for guided missiles. In mid-1916 torpedoes were chosen over bombs for the missile warhead because torpedoes allowed the maximum stand-off distance between the ship and the attacking aircraft. By autumn 1916 a method had been developed and tested to eject the torpedo just before the glider struck the water.

The size of the gliders was then increased significantly since torpedoes were to be carried, and night tests commenced. A flare was added to the glider so it could be seen in the dark and guided to its target. SSW conducted these tests throughout 1917. After more than 75 tests at Biesdorf, Zeppelins

Above: Siemens test glider in 1915.

Above: One of the early larger gliders hanging from the Parseval *P.IV* airship. The configuration is similar to the smaller gliders in the photos on the facing page. The internal guidance package was powered by a battery, and later by a propeller-driven dynamo. Guidance signals were transmitted by thin wires several kilometers long.

Above: As the gliders were enlarged, they were built as biplanes. Successful guided flights against ground targets were made from an airship in 1916.

Above: Experimental glide-bombs in the Biesdorf airship shed in December 1915. Type 1b, the most common, had a wing area of 17.2 ft.^{2} and weighed up to 150 lb. A Type 1c is at far left.

Above: One of the larger test gliders attached to an airship. The guidance system Siemens developed was command guidance by means of wires that unspooled from the glider.

Above: SSW glide-bomb #66 on the launching track at the SSW cable works. In October 1916 the 3.5 m long torpedo was first successfully released above the water of the Spree Canal.

Below: SSW torpedo-glider #2, weighing 300 kg, was successfully dropped from airship *Z.XII* in April 1917 after more than 75 glide-bombs were tested at Biesdorf. The cable spool is between the body and the upper wing.

Above: Two SSW torpedo gliders suspended below an airship, probably Zeppelin *L.35*, which was used for testing.

Below & Right: An Albatros D.III and a 500 kg SSW torpedo glider beneath Zeppelin *L.35* on Jan. 26, 1918. This was the first time an airplane was launched from an airship.

Above & Left: SSW torpedo-glider #7 showing the torpedo in flight mode above and in release mode at left. On August 2, 1918 this glider flew 7.4 km after being released from Zeppelin *L.35* at an altitude of about 1,200 meters. Just after being commanded to turn into the target at about 60 meters altitude, the 7.4 km long guidance wire broke and the glider spun and crashed near the target. This was the last torpedo glider launched from a Zeppelin.

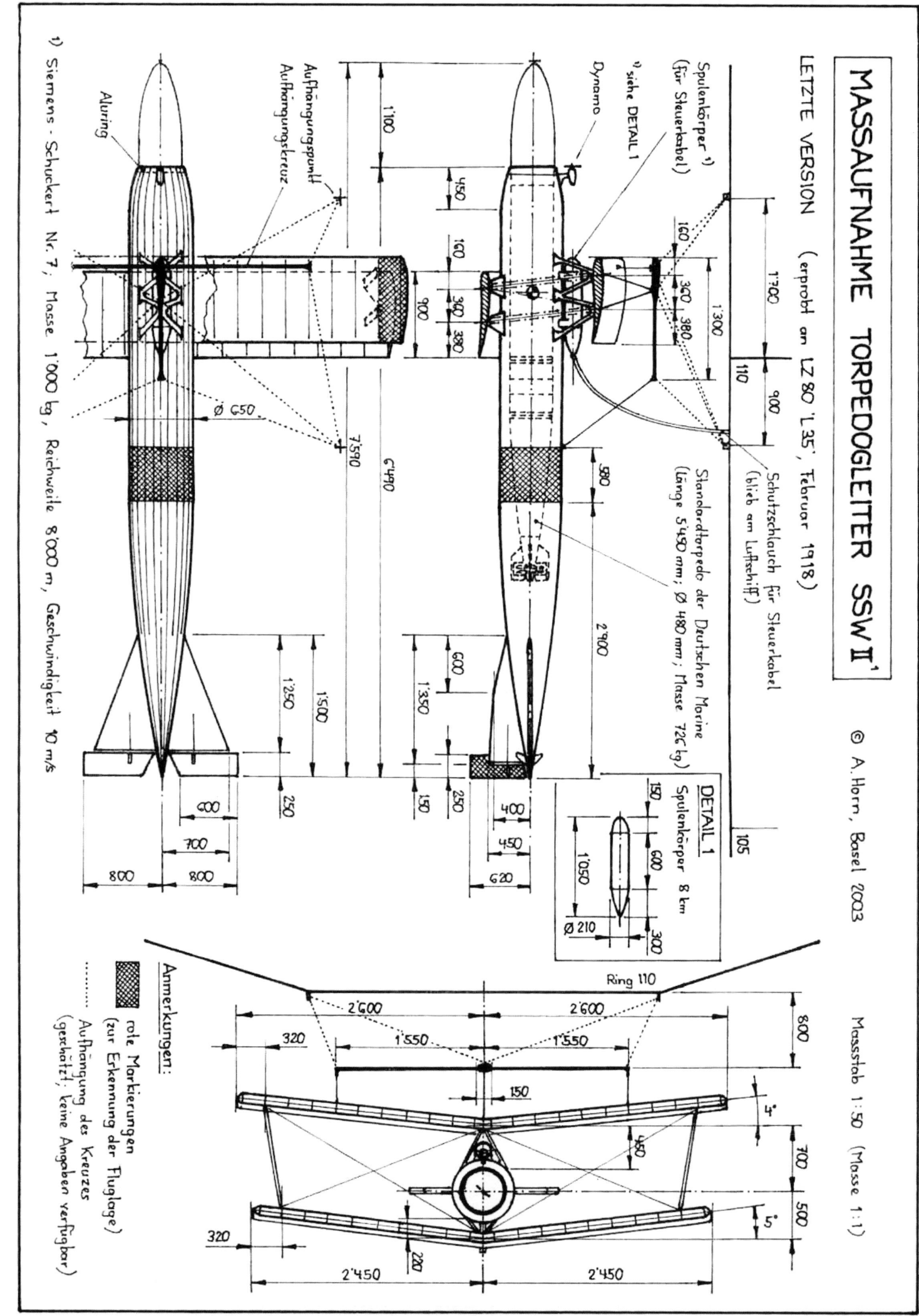

Above: This SSW monoplane torpedo glider weighed 1,000 kg and was developed after the final Zeppelin test in August 1918. The monoplane torpedo glider was intended for use from Zeppelins and, with its lower profile, could also be carried under large bombers. The pod on top of the body carried the spool of guidance wire. Further Zeppelin tests were planned but the Armistice intervened.

were provided to test larger, heavier gliders from higher altitude. From summer 1917 gliders weighing 660 pounds (300 kg) were launched from airship *Z.XII* in Hannover. In Autumn gliders of 1,100 pounds (500 kg) and 2,200 pounds (1,000 kg) were tested from Zeppelin *L.35* at Jüterbog and *PL 25* at Potsdam.

After the older airships were dismantled *L.35* was seldom available for the glider tests, and when a glider collapsed while suspended below *L.35* in April 1918, further tests were suspended until the reason could be determined and fixed. Tests resumed in summer and the last test launch of a torpedo glider was made August 2, 1918. *L.35* released the 1,000 kg biplane torpedo glider from 4,000 feet altitude at a target 4.6 miles (7.4 km) away. At an altitude of about 200 feet above the target and just after being commanded to turn into the target, the 5-mile long

wire broke and the glider spun and crashed.

Meanwhile, SSW had designed monoplane torpedo gliders that would fit under the wings of bombers. However, the Armistice intervened before any of these could be test-launched and work on the torpedo-glider project was ended in December 1918.

Above: This advanced monoplane torpedo glider had a dynamo in the tail to power the guidance system and the control wires ran out through the hollow dynamo shaft.

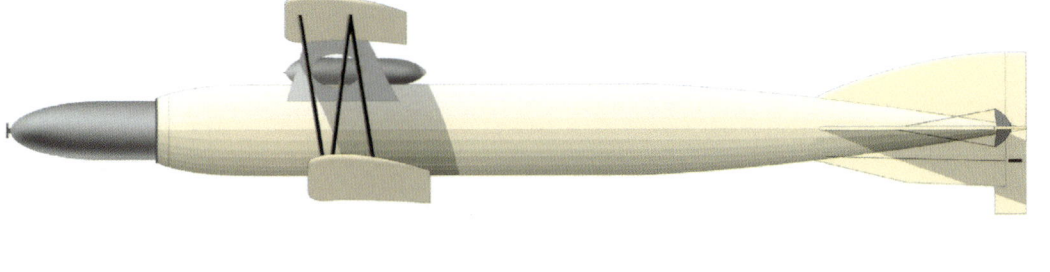

SSW Biplane Torpedo Glider Guided Missile

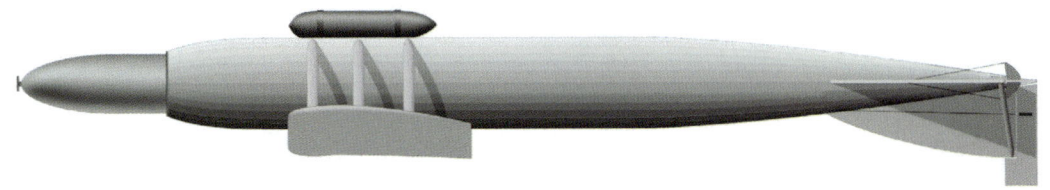

SSW Monoplane Torpedo Glider Guided Missile

Siemens-Halske Counter-Rotary Engines

By Dick Bennett

The development and history of Siemens & Halske's rotary engines were closely tied to the fighter aircraft built by that division's sister company, Siemens-Schuckert. These engines were intended primarily to power SSW's fighters, and, as it turned out, the fate of one clearly influenced the fortunes of the other.

In 1912, Siemens began preliminary studies of aircraft and their power requirements and concluded that rotary engines, with their compact structure, high power-to-weight ratios and smooth, low-vibration operation were the optimum powerplants for aircraft. However, they also found existing designs suffered from high fuel and oil consumption and an asymmetric gyroscopic effect on aircraft maneuverability.

Another shortcoming soon became apparent – concerns about material strength limitations constrained rotaries to a maximum rotational speed of about 1,200 RPM. Although the structural concerns may have been exaggerated, high rotational speeds had another disadvantage – the frictional air resistance to a rotary spinning at 1,200 RPM could consume 10 to 12% of its power output, and this loss increased with the cube of the engine speed. This put rotaries at a disadvantage against stationary engines that could run at speeds of 1,750–2,100 RPM, producing much higher power from a given displacement. Siemens chose to build a rotary engine that could run at comparable crankshaft speeds, specifically, 1,800 RPM.

To withstand the stresses on rotary engine spinning at 1,800 RPM, its structure would have to be beefed up, adding unwanted weight. Even if the added power justified the increased weight, the engine's gyroscopic moment would increase, as would air resistance to the spinning cylinders, consuming significant power. To get higher crankshaft speeds, some other technique had to be found.

That technique was counter-rotation, achieved by placing a differential gearbox between the engine and propeller, using the gearbox case as the engine's main mounting point. Crankshaft and crankcase were both now free to rotate in opposite directions. The crankshaft turned 1,800 RPM counterclockwise (viewed from the propeller end), but due to a 2:1 gearing ratio, the crankcase, cylinders and propeller were driven at 900 RPM in the opposite direction. This reduced the net propeller speed to 900 RPM counterclockwise, achieving higher propeller efficiency, lower gyroscopic moments, and lower air resistance while pulling more horsepower from the engine without overstressing it.

Siemens' first rotary was the Sh.0, which was submitted to *Idflieg* in October 1915. It was a nine-cylinder engine, clearly patterned on the French Gnome *Monosoupape*, but mated to a gearbox that provided counter-rotation. It met Siemens' performance objectives, but also retained the shortcomings of the Gnome valve system.

The Gnome's valve gear was derived from a single-cylinder stationary engine (licensed, ironically, from Oberursel in Germany) with an automatic inlet valve situated in the piston crown. This was held closed by a spring, but on the piston downstroke, the differential pressure between the combustion chamber and the outside atmosphere overpowered the spring, forcing the valve open. However, as altitude increased and air density fell off, the differential pressure decreased, and the valve wouldn't open as far. This, combined with the lower air density at altitude, caused power to fall off quickly. To make matters worse, oil varnish deposits could build up around the valve, causing it to stick. Siemens improved on this by replacing the main valve spring with a mechanical drive. A cam on the connecting rod's wrist pin boss, working through a lever, held the valve tightly closed on the upstroke and then moved away to let the valve open on the

Siemens-Halske Counter-Rotary Engine Summary						
Model	# Cylinders	Power	Bore	Stroke	Displ.	Notes
Sh.0	9	90 hp	100 mm	130 mm	9.19 L	
Sh.I	9	110 hp	114 mm	130 mm	11.94 L	
Sh.Ia	9	125 hp	114 mm	130 mm	11.94 L	Over-compressed Sh.I, aka Sh.II
Sh.III	11	205 hp	124 mm	140 mm	18.60 L	Initial production engine
Sh.IIIa	11	205 hp	124 mm	140 mm	18.60 L	Final production engine

downstroke. A pair of counterweights helped reduce operating forces but did not actually open or close the valve.

The Sh.0 didn't find favor with *Idflieg* – by the time most of its teething problems had been rectified, its 90 hp rating was no longer considered sufficient for the new generation of military aircraft. It was fitted to only eight SSW E.I fighters and a few prototypes. By August 1916, when the first of these Siemens *Eindeckers* left the factory, they were no longer suitable for front-line service, and most went to flying schools.

The Siemens monoplane wasn't alone; all the German *Eindeckers* had been outclassed by the French Nieuport 11 and 16 fighters that appeared early in 1916. *Idflieg* requested that several manufacturers build biplane or sesquiplane fighters to counter them. Siemens-Schuckert took the simplest and most direct route by building the D.I, a copy of the Nieuport. At the same time, S & H was working on an up-rated version of the Sh.0 to power the new generation of fighters. This materialized as the Sh.I, essentially an Sh.0 with a larger cylinder bore and a standard power rating of 110 hp. It became the D.I's powerplant.

Idflieg had high hopes for the SSW D.I, but by the time it was finally in production, the Nieuports it was based on had been serving a year and were themselves becoming obsolescent. An initial order of 250 D.Is was cut to 95, and of these, only a small number made their way to front-line units. The balance of the order, also fitted with Sh.I engines, was shipped to flying schools and *Idflieg's* test establishment. A higher compression Sh.I, designated Sh.Ia by the factory, but also referred to unofficially as the Sh.II, and rated 125 hp, was fitted to four experimental aircraft – two D.Ib, a Dr.I and the odd push-pull DDr.I, but none of these aircraft went into production. With that, the Sh.I's career came to an end.

In spite of its lack of success, the Sh.I proved Siemens' design theories were correct; it produced 9.66 hp per liter of displacement, compared to 6.76 for the 110 hp Gnome 9B-2 and 7.33 for the Le Rhône 9Ja, its closest competitors. Even the added weight of the gearbox didn't put it at a disadvantage with its French counterparts. The Sh.I produced 0.82 hp per kilogram of weight, the Gnome, 0.81 and the Le Rhône, 0.75.

The performance race between Allied and Central Powers aircraft continued. More speed and faster climb rates were required, and the added weight of the now-standard two machine guns, their ammunition, and other equipment demanded higher-powered engines. Hispano-Suiza's 150 hp 8A and Mercedes' D.III, good for 160 hp, now set the standard. Siemens-Halske's response was to build

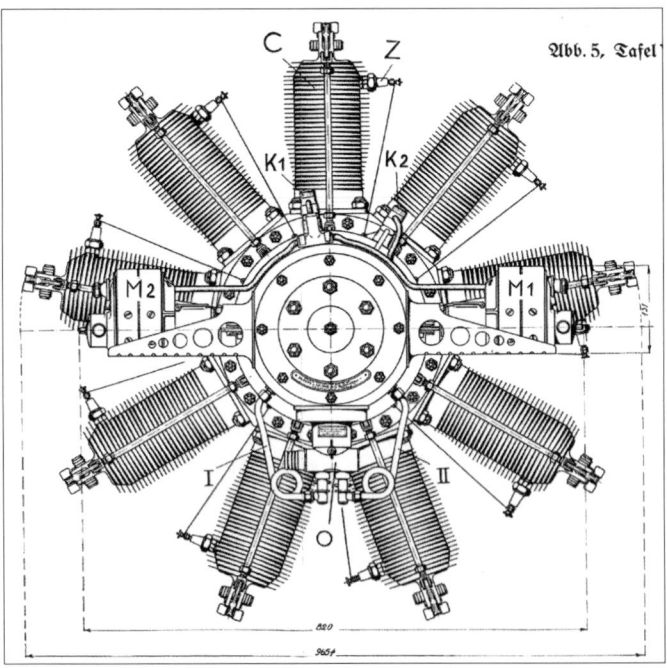

Above: Front view of Sh.I engine.

Left: Three-quarter rear view of Sh.I engine.

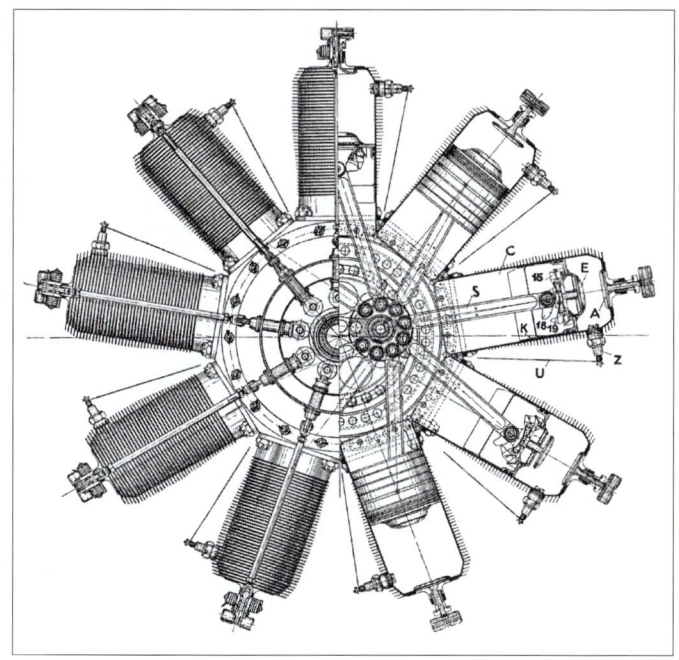

Above: Cross section of Sh.I engine.

Above: Gear section of Sh.I engine.

Below: Installing an Sh.I engine into a Fokker prototype.

Below: Sh.I engine after installation in a Fokker prototype.

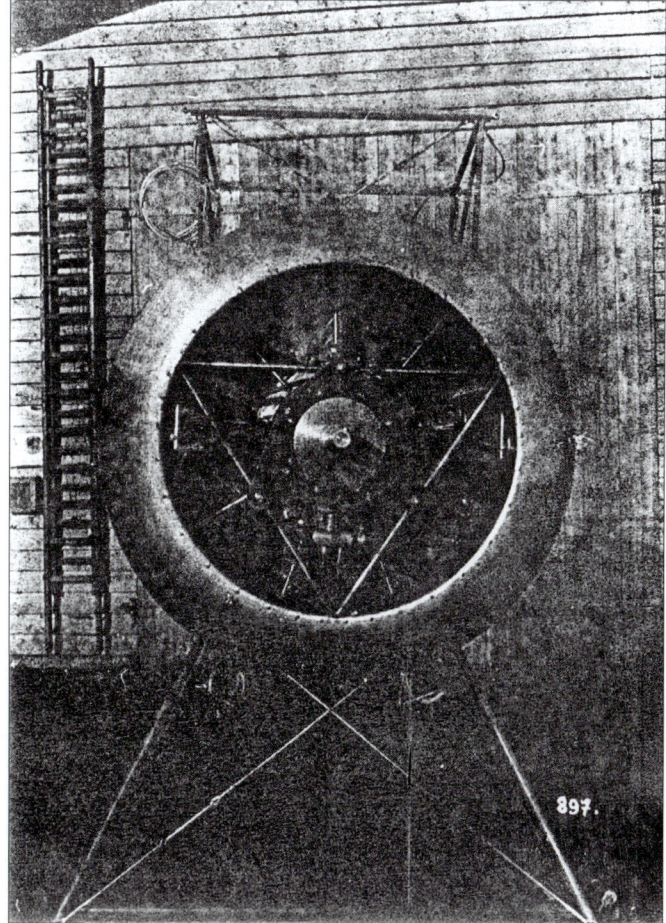

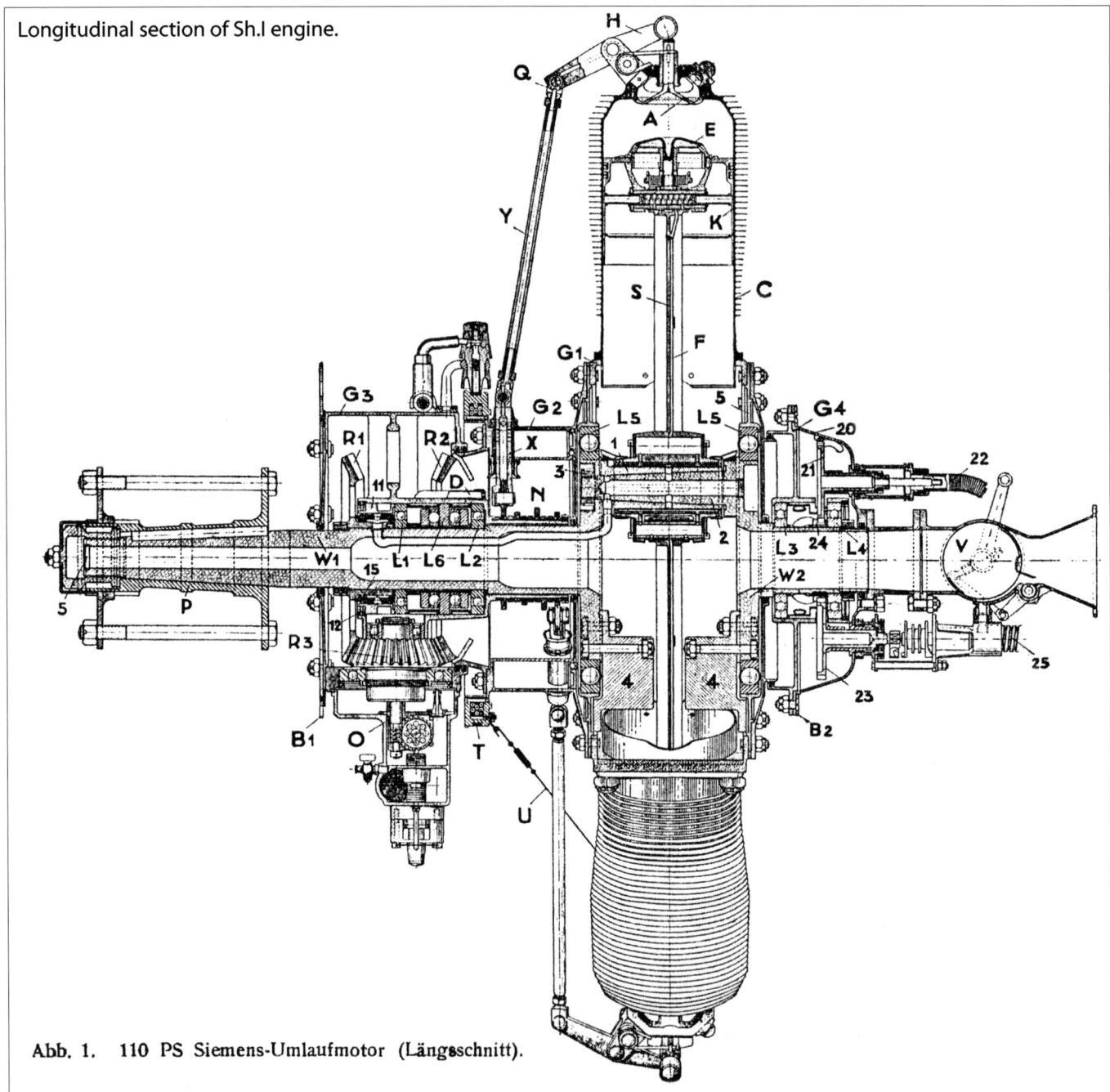

Longitudinal section of Sh.I engine.

Abb. 1.   110 PS Siemens-Umlaufmotor (Längsschnitt).

an 18-cylinder, two-row rotary making maximum use of Sh.I components. This engine was expected to produce 220 hp, but was so heavy and complex that Siemens abandoned it early in development.

Starting with a clean sheet of paper, they created what became the Sh.III. It materialized as a massive-looking 11-cylinder, single-row rotary. Inlet and exhaust valves, located in the cylinder heads, were operated by pushrods actuated by cam rings in a housing at the front of the engine. The gearbox was moved to the rear of the engine, where a train of four bevel gears, reminiscent of an automobile

differential, translated the crankshaft's rotation to the crankcase/propeller assembly, driving it in the opposite direction. Ready for testing by June 1917, the engine immediately showed its promise, producing 160 hp. With continued refinement, it would eventually produce over 200 hp at sea level. Construction materials were fairly conventional – all major components were machined from steel, except the pistons, which were cast aluminum.

At 5.1 to 1 compression, the Sh.III was considered a high compression engine. By today's standards, that seems laughably low, but this was a time of low

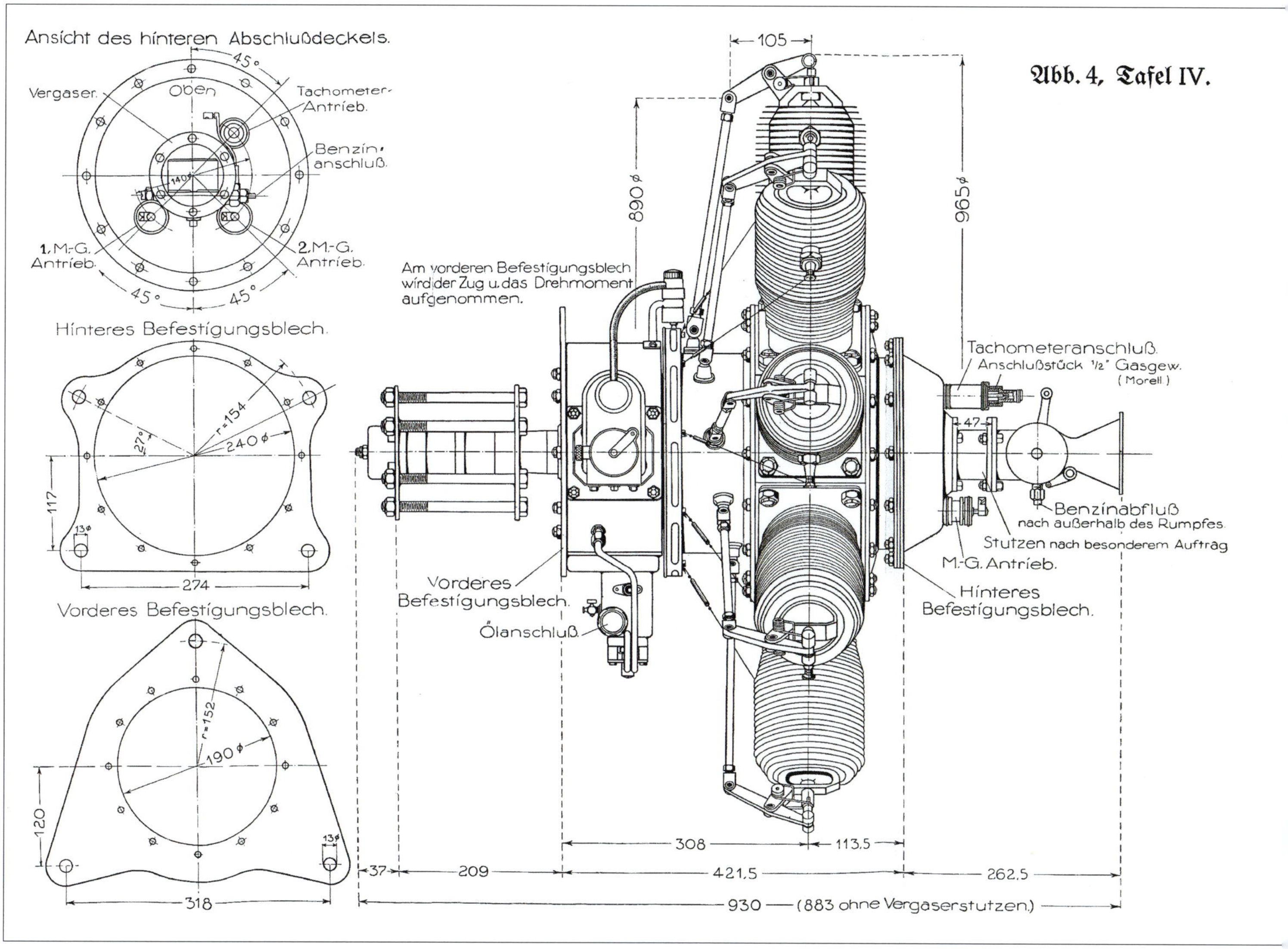

Abb. 4, Tafel IV.
Ansicht des hinteren Abschlußdeckels.
45°
Oben
Vergaser.
Tachometer-Antrieb.
Benzin-anschluß.
140
1. M.-G. Antrieb.
2. M.-G. Antrieb.
45°
45°
Hinteres Befestigungsblech.
r=154
240 ø
27°
117
13 ø
274
Vorderes Befestigungsblech.
r=152
190 ø
120
13 ø
318
105
890 ø
965 ø
Am vorderen Befestigungsblech wird der Zug u. das Drehmoment aufgenommen.
Tachometeranschluß.
Anschlußstück 1/2" Gasgew. (Morell)
47
Benzinabfluß nach außerhalb des Rumpfes.
Stutzen nach besonderem Auftrag
M.-G. Antrieb.
Hinteres Befestigungsblech.
Vorderes Befestigungsblech.
Ölanschluß.
308
113,5
37
209
421,5
262,5
930 (883 ohne Vergaserstutzen.)

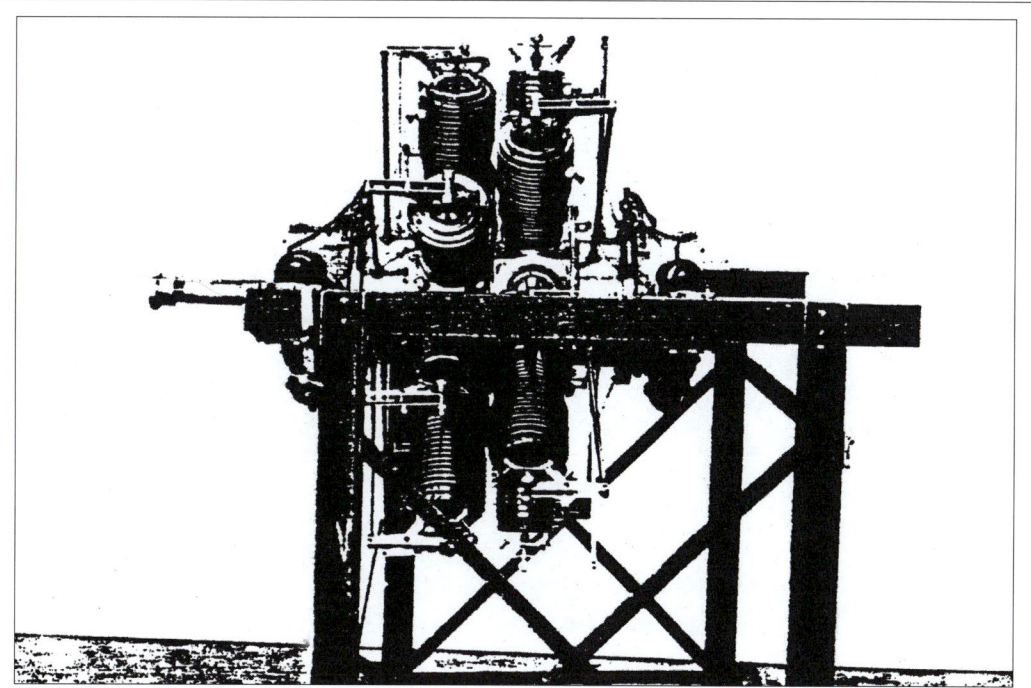

Above: Sh.I engine installation.

Right: Side view of the experimental 18-cylinder, two-row engine on a test stand.

Below: Front view of the experimental 18-cylinder, two-row engine.

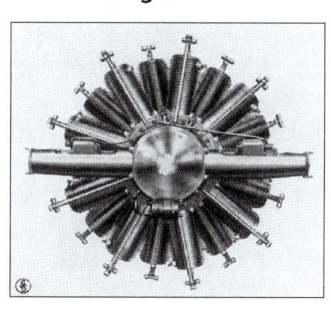

octane fuels and aero-engine compression ratios that typically ranged from 4.5 to 5.5.

The Sh.III's performance and potential for high altitude use caught the attention of *Idflieg*, and the clincher was the spectacular climbing performance of an Sh.III-powered SSW D.III prototype in September 1917. *Idflieg* immediately gave the go-ahead for a limited production run of this promising new engine, even though it hadn't passed the customary endurance test. Eventually, Siemens-Halske received an order for 2,000 units, and later, a manufacturing license and order for 1,000 more engines was placed with Rhemag (Rhenania Motorenfabrik A. G.).

On the basis of the January 1918 Fighter Trials, *Idflieg* ordered 50 SSW D.III and 50 Pfalz D.VIII fighters along with an evaluation batch of Pfalz D.VII or D.VIII biplanes, all powered by the Sh.III. The first SSWs were sent for evaluation by *Jagdgeschwadern* II and III in March and April 1918. After a few hours' operation, the engines began seizing, with pistons, rings, valves and other components fracturing. Production of the engines and SSW D.IIIs was stopped; surviving D.IIIs at the front were returned to the factory.

The search for a cause of the failures produced plenty of suspects – inadequate cooling, the Voltol castor oil substitute used by the engines, component design, and metallurgy. If time were not an issue, these probably would have been evaluated individually and the innocent crossed off the list. But this was a crisis – Germany was at war and needed all the high performance engines and fighters it could get. Everything was "fixed", whether necessary or not.

The most important changes involved engine cooling, and this was obvious when the reworked aircraft and engines were reintroduced to combat units in July. The bottom half of the cowling had been cut away, scoops had been added to the propeller spinner to direct air onto the crankcase, and an additional exhaust duct had been shoehorned into the front of the fuselage, its outlet discharging on the right side of the aircraft. The Voltol lubricant, an electrolyzed blend of mineral and rapeseed oils, was still regarded with suspicion, but no alternatives were available. Internally, the pistons were redesigned. The original pistons had a rather deep, square-shouldered recess in the crowns, and designers apparently suspected the sharp, drastic section changes in the castings might be causing too much stress. Consequently, they redesigned them with a shallower dished recess in the crown, much like the Le Rhône.

Left: Front Front view of the Siemens-Halske Sh.III. The eleven-cylinder engine had a bore of 124mm, a stroke of 140mm, and a displacement of 18.60 liters. Originally delivering 160 hp, it was gradually developed to deliver 205 hp. After the shortcomings of the original design were corrected, the Sh.IIIa produced by Siemens-Halske and their licensee, Rhemag, proved to be very reliable. Fighters using the Sh.III/IIIa had good speed and an exceptional rate of climb.

This created another problem. Because the piston recess was part of the combustion chamber, reducing its volume caused a corresponding reduction in combustion chamber volume, and with it, power, but it increased the compression ratio on an engine already considered too highly stressed. Siemens' solution to the problem was elegantly simple – they restored the original combustion chamber volume and compression ratio by making the cylinders 10 mm taller. This allowed existing crankshafts and crankcases to be re-used. Engine diameter increased from 1,050 to 1,070 mm, but the only other visible difference was the addition of two fins to the cylinders (the fins were spaced 5mm apart).

The modified engine passed *Idflieg's* endurance test in June 1918 and was released for production as the Sh.IIIa, to emphasize it was not the trouble-plagued original. Contrary to some descriptions, it was not a higher compression or more powerful version of the Sh.III – all the changes simply maintained its original performance.

Sh.IIIa engines from Siemens-Halske and Rhemag were used primarily on SSW's D.III and D.IV fighters, in addition to the Pfalz D.VIIIs and Dr.Is ordered after the First Fighter Trials. Individual engines were

Siemens Poem

Ten little Siemens were circling around a barn,
A piston ring froze up tight, nine were left without harm.
Nine little Siemens were playing out to join the hunt,
The spark plugs flew away from one, eight are left to hunt.
Eight little Siemens, they flew up very high,
A piston went out fast, now only seven are all right.
Seven little Siemens were flying on top of a spot,
The magneto didn't turn on one, only six are left on lot.
From six little Siemens flew out one with red socks,
Connecting rod salad right away, only five came back to box.
Five little Siemens we still have left to fly,
A throttle valve failed, but four still take the sky.
Four little Siemens were flying high and free,
a cylinder turned blue, we now have only three.
Three little Siemens are going very fast,
The bearnings blow out quickly, only two who pass the test.
Two little Siemens, flying lonely like father and son,
An engine seizes completely tight, and now there's only one.
One little Siemens went up in brave salute,
He came up only 3,000 meters and now too is Kaputt!

Translated from the original German by Rick Duiven.

Right: Rear view of the Siemens-Halske Sh.III eleven-cylinder counter-rotary engine. The low rotational speed of the cylinders, 900 RPM compared to 1200–1300 RPM for conventional rotary engines, reduced the cooling air flow over the cylinders, which reduced cooling drag but also made cooling marginal. Problems with insufficient cooling caused a number of engine changes as well as airframe changes to the fighters using this engine to increase cooling air flow over the cylinders. These airframe changes were especially noticeable in the SSW D.III and D.IV fighters, which had a number of engine cowling modifications as well as the addition of cooling air scoops to the propeller spinner.

also installed in numerous prototypes – Albatros D.XI, Fokker V.7S, V.13-II and V.28, Pfalz D.VII and D.X, SSW D.V and D.VI, and Roland D.IX and D.XVI – that didn't make the cut.

Sources disagree on the numbers, but by war's end, Siemens-Halske had produced about 550 Sh.IIIs, and Rhemag produced 258 to 289.

The Sh.III's horsepower has long been a source of discussion, because various sources quote different figures – 160, 200, 220 and 240 hp, for example. Part of the confusion seems to stem from the fact that, as a high altitude engine, the Sh.III was expected to deliver a surplus of power at sea level to insure reasonable performance at altitude. The most definitive figure appears to come from a test of a war booty Sh.IIIa by the British Air Ministry in 1920. It produced a maximum of 207 hp at 1700 RPM (850 propeller RPM). At its maximum speed of 1800 (900) RPM, power fell off slightly to 204 hp, suggesting the engine had exceeded its breathing capacity.

How did the Sh.IIIa stack up against its closest competitors, the 160 hp Gnome 9N and the Bentley B.R.2? In terms of power per unit weight of engine, the three were virtually identical, but the Sh.3a outscored its competitors in horsepower per unit of displacement – 11.0 hp per liter for the Sh.IIIa, 10.06 for the Gnome, and 9.54 for the Bentley.

For an engine built in relatively low numbers, the Sh.III proved to be a hardy survivor. Because of its unique features, the victorious Allies seized numerous examples for technical evaluation, and Sweden even purchased 10 of them for possible use

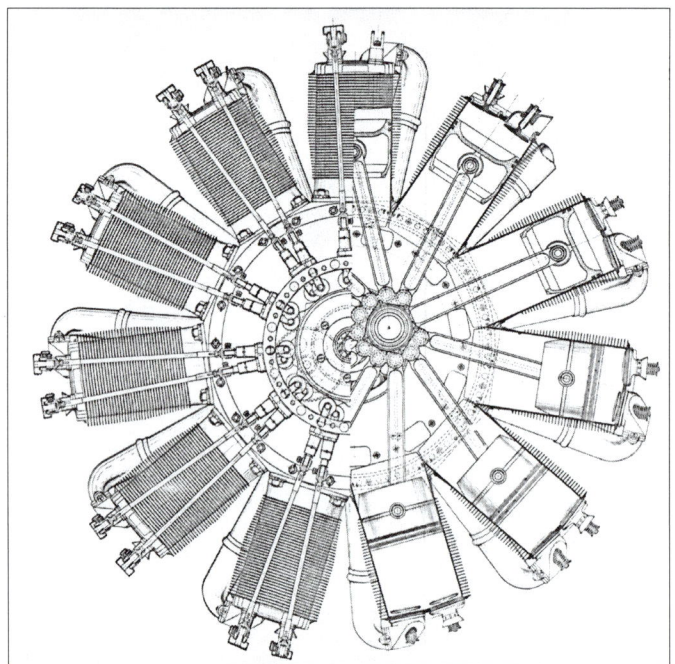

Above: Cross section of Sh.III engine.

by its air force. Today, survivors exist in Germany, Austria, Poland, the U. K., Sweden, Switzerland, and the U. S., and one Finnish museum displays components of a torn-down Sh.III.

Siemens-Halske did not produce enough counter-rotary engines to challenge the dominance of their competitors, and the Sh.III doesn't have the distinction of being the most powerful rotary built – that honor goes to the Bentley B.R.2 – but it was certainly the most mechanically sophisticated rotary design. As such, it represents the peak of rotary engine development.

Below: Longitudinal section of Sh.III engine.

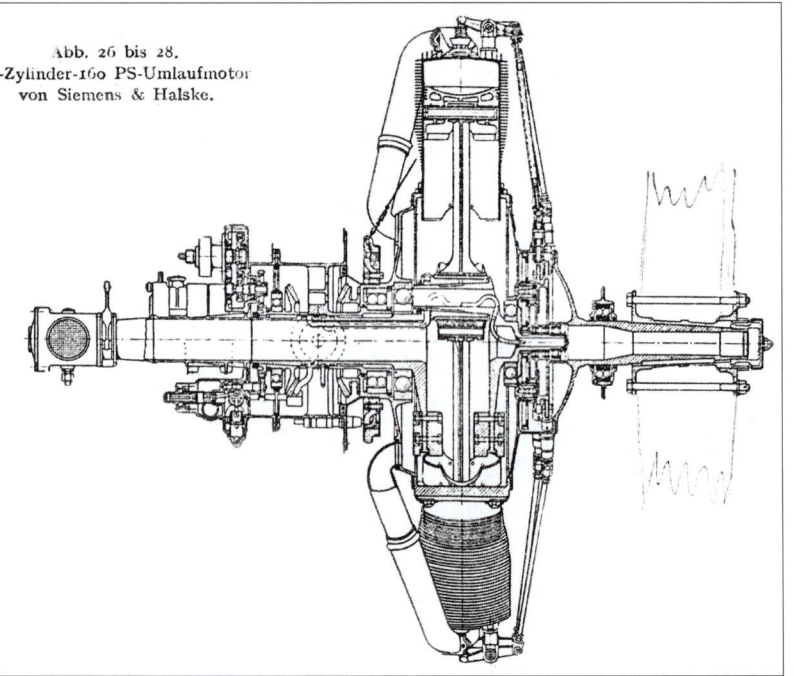

Below: Installation sketch of Sh.IIIa engine.

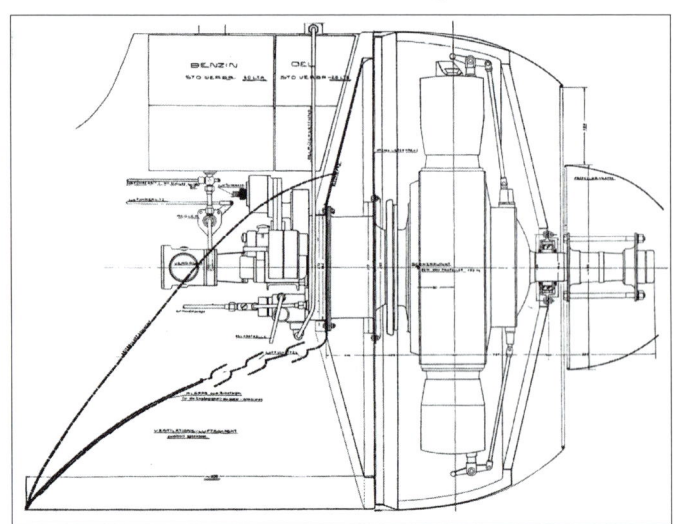

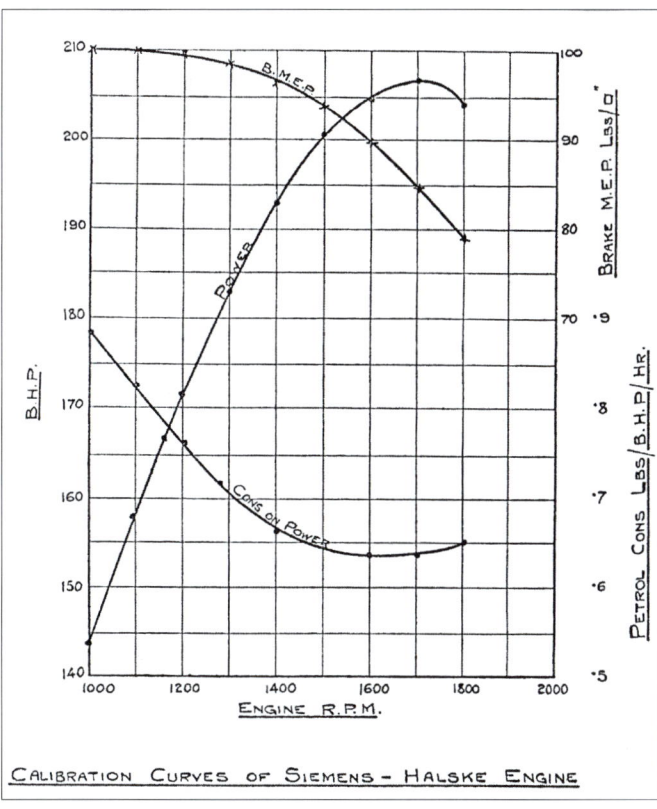

CALIBRATION CURVES OF SIEMENS - HALSKE ENGINE

Above: Power curves for the Sh.IIIa engine from a British Air Ministry test done in 1920. Maximum power developed was 207 hp at 1,700 RPM; at 1,800 RPM the power dropped off slightly to 204 hp.

Above & Below: The Siemens-Schuckert non-rigid airship SSL in its hangar (above) and in low-level flight (below).

Below: The Siemens-Schuckert non-rigid airship SSL in flight, probably 1911.

In Retrospect

Siemens-Schuckert was a huge company with enormous engineering resources that used them to make an important contribution to German aviation during WWI. Despite its engineering talent, SSW got off to a disappointing pre-war start in aviation and dropped aeronautics until the war started and Germany desperately needed airplanes.

SSW's initial move back into aviation was getting involved in R-plane construction. The Forssman design was available and SSW started with it. Reliable production engines were used, but the Forssman airframe had fundamental structural and aerodynamic deficiencies that could not be overcome despite massive investment of engineering resources and numerous rebuilds; development of this doomed design should have been abandoned much earlier.

In parallel SSW engaged the Steffen brothers as designers and progressed with their original R-plane designs. Despite their unorthodox appearance the SSW R-planes had fundamentally good airframes with good handling qualities. Unfortunately, the design used a central power system to facilitate in-flight maintenance and avoid excessive drag of a windmilling propeller in case of engine failure. This meant that installing different engine types required significant engineering effort and rebuilding part of the central fuselage structure. By far the worst problem was the failure of the Maybach HS engines chosen to power the aircraft. The failure of the HS engine greatly impeded the SSW R-plane program with the result that they appeared at the front much later than planned. The SSW R-planes were modestly successful on operations but were quickly superseded by more advanced designs from the Zeppelin-Staaken company. The next generation of SSW R-planes and the L.I were too late for the war.

Soon after SSW launched their R-plane program they also started building fighters. The first fighter designs from SSW were derivative and made no impact. The SSW D.I, based on the Nieuport airframe and powered by a Siemens-Halske engine, was also derivative and built only in small numbers. However, the engineering and production experience SSW gained with the D.I and its Sh.I engine enabled SSW to develop a new generation of original fighter designs powered by the more powerful Sh.III engine from Siemens-Halske, SSW's sister company.

The SSW D.III and D.IV were the new-generation SSW fighters that reached operational service. Both were plagued by engine teething problems, and after a brief period at the front in Spring 1918 had to be returned to SSW for engine replacement. Additional engine and airframe development were needed to achieve the required reliability, then the D.III and D.IV returned to the front.

Once in combat using reliable engines, the D.III and D.IV proved to be very maneuverable fighters with exceptional climb and ceiling. Not as fast as their Spad and S.E.5a adversaries, the SSW fighters were nevertheless considered superior by their pilots. In fact, many SSW pilots considered their mounts to be the best fighters of the war.

Of course, this opinion conflicts with the more general opinion that the BMW-powered Fokker D.VII was the war's best fighter. Although the SSW fighters had better climb and ceiling, the Fokker with BMW engine was faster, more reliable, and much easier to fly. Moreover, the Fokker arrived in quantity at the front sooner and was built in great numbers, its fine qualities making it the standard of comparison. Engine development problems impeded the SSW fighters arrival at the front, so they were relatively few in number and made less of an operational impact despite their excellence.

It was unfortunate for SSW, and fortunate for Allied pilots, that the SSW D.VI parasol monoplane was too late for combat. The D.VI inherited the exceptional climb, ceiling, and maneuverability of the D.III and D.IV biplanes while being considerably faster. In fact, the climb and ceiling of the SSW fighters made them reasonable alternatives to the troubled Rumpler D.I that Rumpler was never able to develop into a practical fighter.

In addition to aircraft and the Siemens-Halske series of rotary engines, SSW also aggressively developed wire-guided glide bombs for attacking surface targets, especially warships, for which a torpedo warhead was used. These were too late for operations during the war, but Germany employed anti-ship missiles in WW2.

Finally, SSW produced nearly 180 aircraft under license, bringing its total to almost 600 aircraft.

Although innovative in engines and missiles, SSW built its aircraft using the wood, wire, and fabric construction typical for the time, and their aerodynamics were also conventional for the period. Their later fighter designs were excellent based largely on the superior performance of their Siemens-Halske Sh.IIIa engine and good airframes, but SSW made no fundamental structural or aerodynamic advances like competitors Junkers, Fokker, or Zeppelin-Lindau.

Bibliography

Books

Franks, Norman, Bailey, Frank, and Guest, Russel, *Above the Lines*, Grub Street, 1993.

Franks, Norman, Bailey, Frank, and Duiven, Rick, *The Jasta Pilots*, Grub Street, 1996.

Gray, Peter, and Thetford, Owen, *German Aircraft of the First World War*, second revised edition, New York: Doubleday & Company, Inc., 1970.

Green, William, and Swanborough, Gordon, *The Complete Book of Fighters*, Smithmark Publishers, 1994.

Gray, Peter, and Thetford, Owen, *German Aircraft of the First World War*, second revised edition, New York: Doubleday & Company, Inc., 1970.

Grosz, Peter M., *SSW D.III–D.IV*, Albatros Publications, Berkhamstead, 1991.

Grosz, Peter M., *The SSW R.I*, Albatros Publications, Berkhamstead, 2001.

Haddow, George W., and Grosz, Peter M., *The German Giants: The German R-Planes 1914–1918*, third edition, Putnam, London, 1988.

Herris, Jack, *Germany's Fighter Competitions of 1918*, Aeronaut Books, 2013.

Herris, Jack, *Germany's Triplane Craze*, Aeronaut Books, 2013.

Articles

Grosz, Peter M., "Frontbestand" *WW1 Aero* No.107, Dec. 1985, p.60–66.

Grosz, Peter M., "Frontbestand" *WW1 Aero* No.108, Feb. 1986, p.66–69.

Kruger, Egon, and Grosz, Peter, "A History of Siemens-Schuckert Aircraft and Missiles, 1907–1919", p.193–229, *Cross & Cockade Journal*, Volume 13, Number 3, Autumn 1972.

Internet

(1) http://en.wikipedia.org/wiki/Siemens
(2) http://www.siemens.com/history/en

Index

SSW E.I

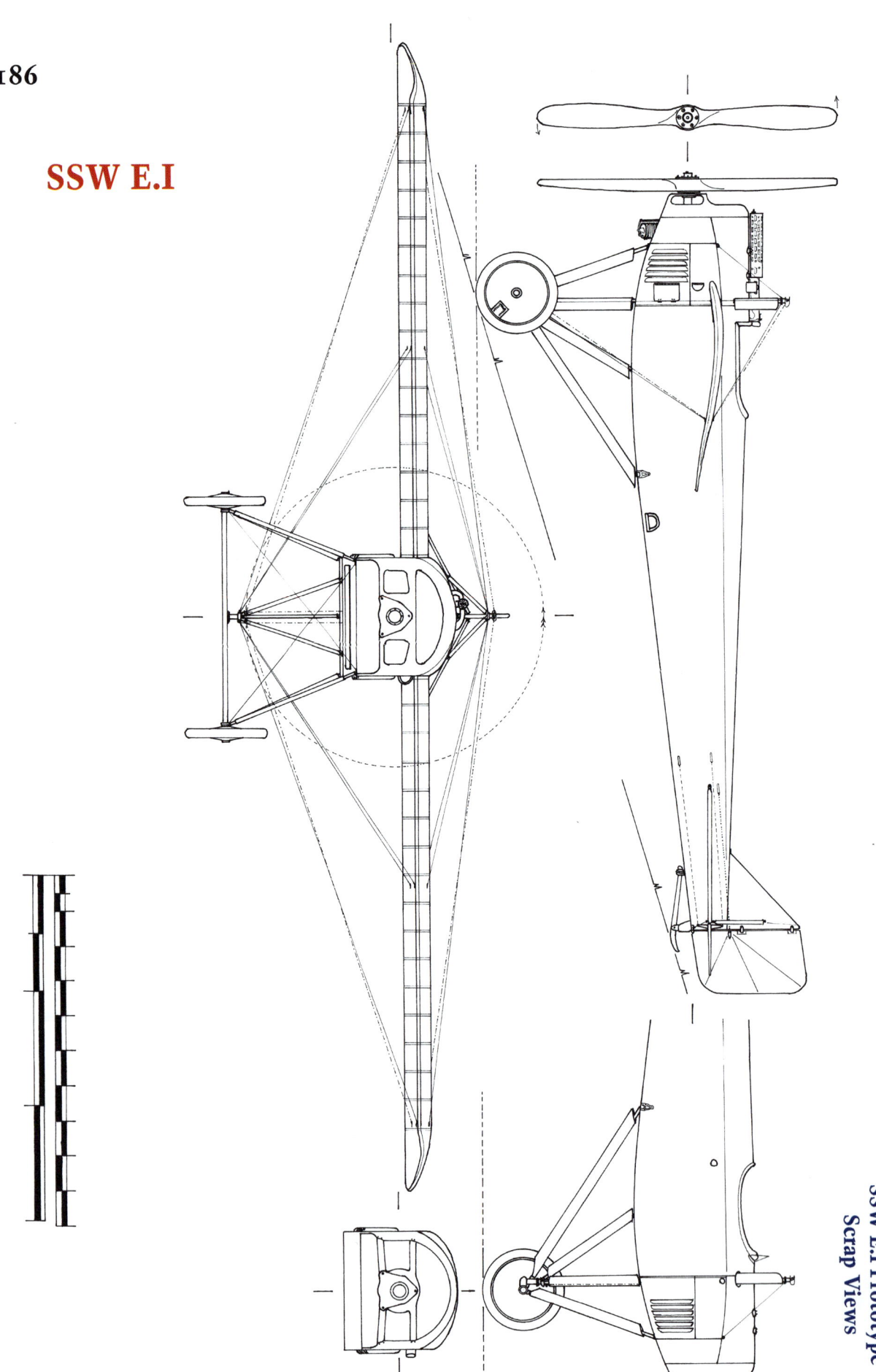

SSW E.I Prototype
Scrap Views

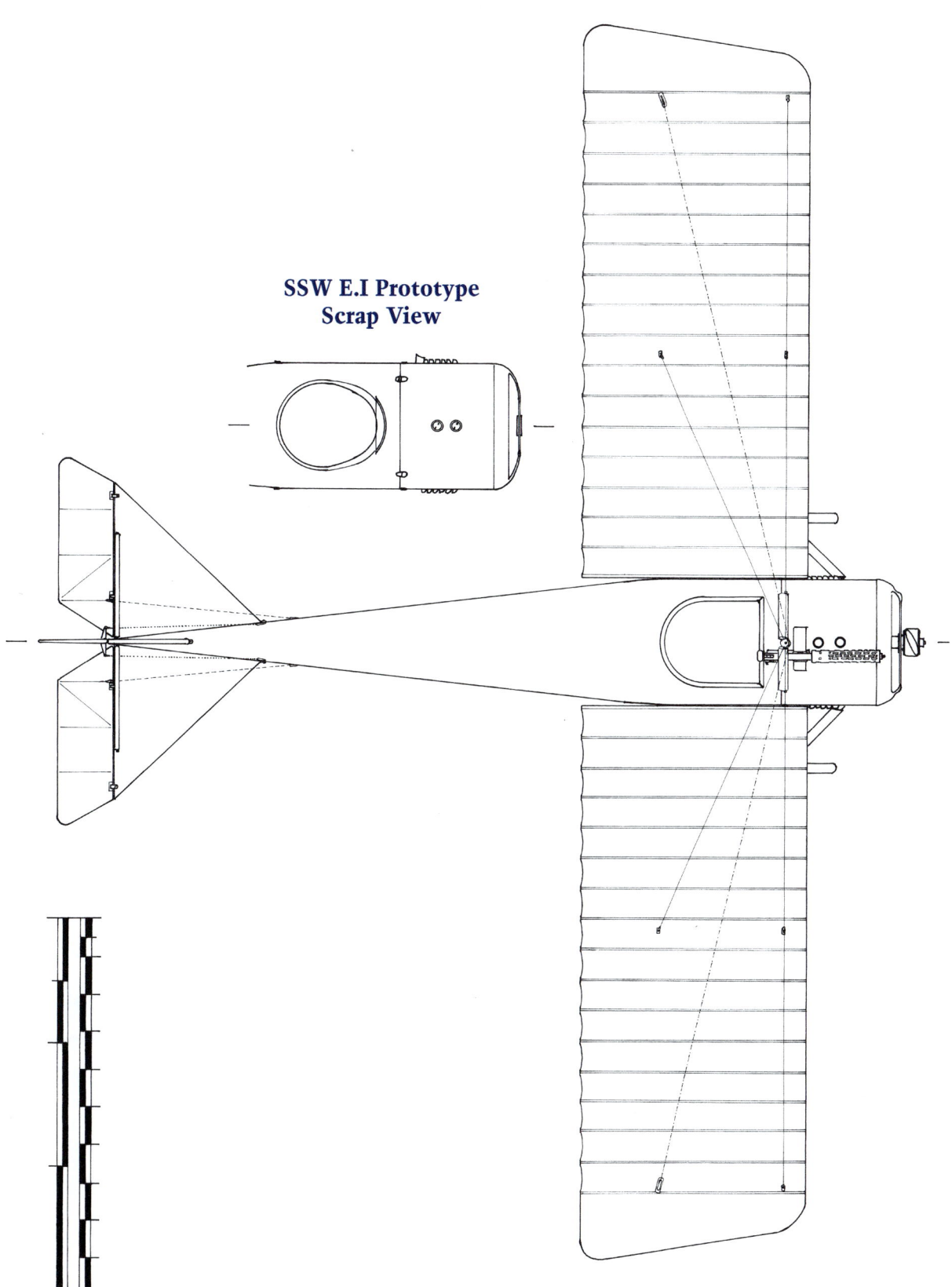

SSW E.I Prototype
Scrap View

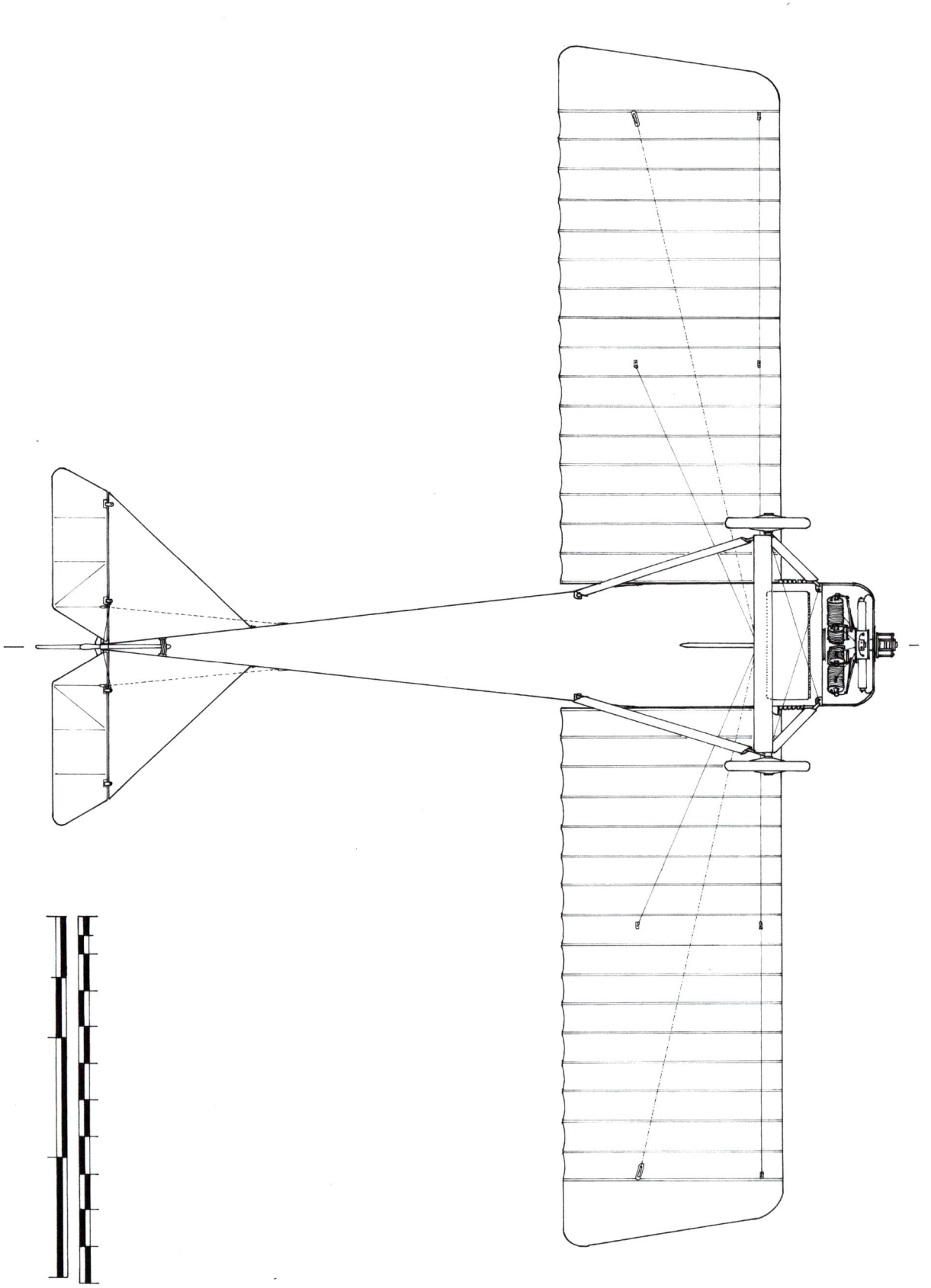

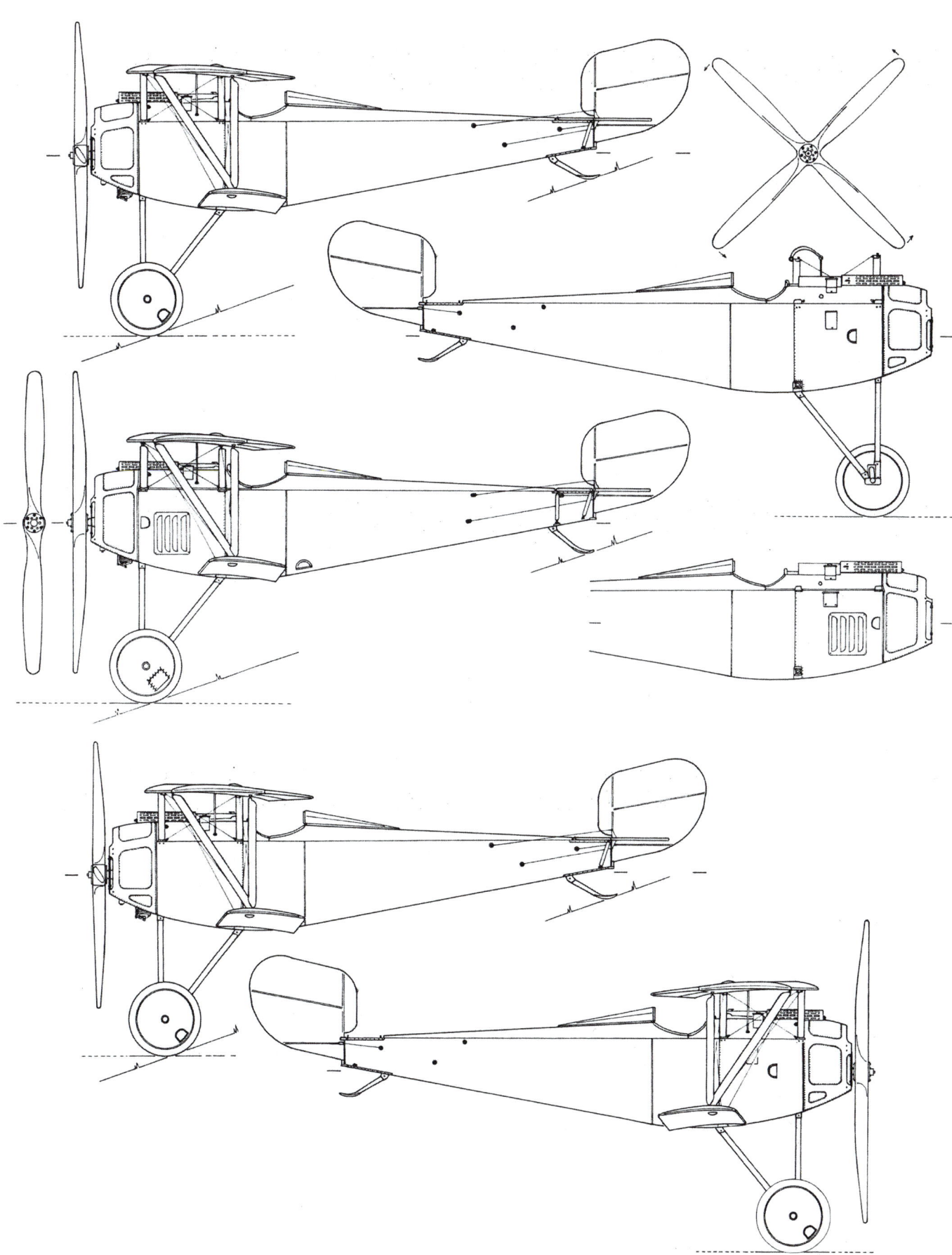

SSW D.I Prototype

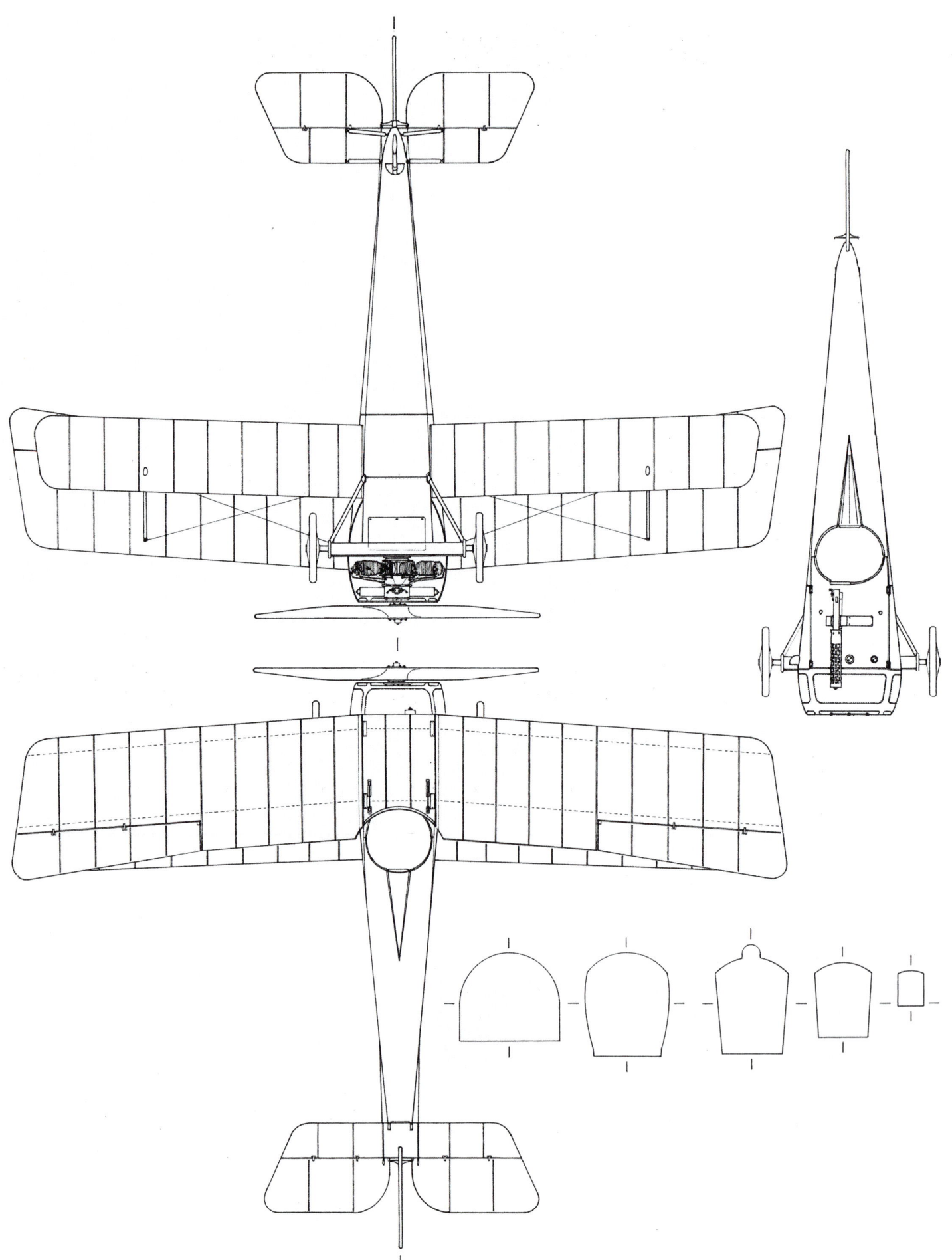

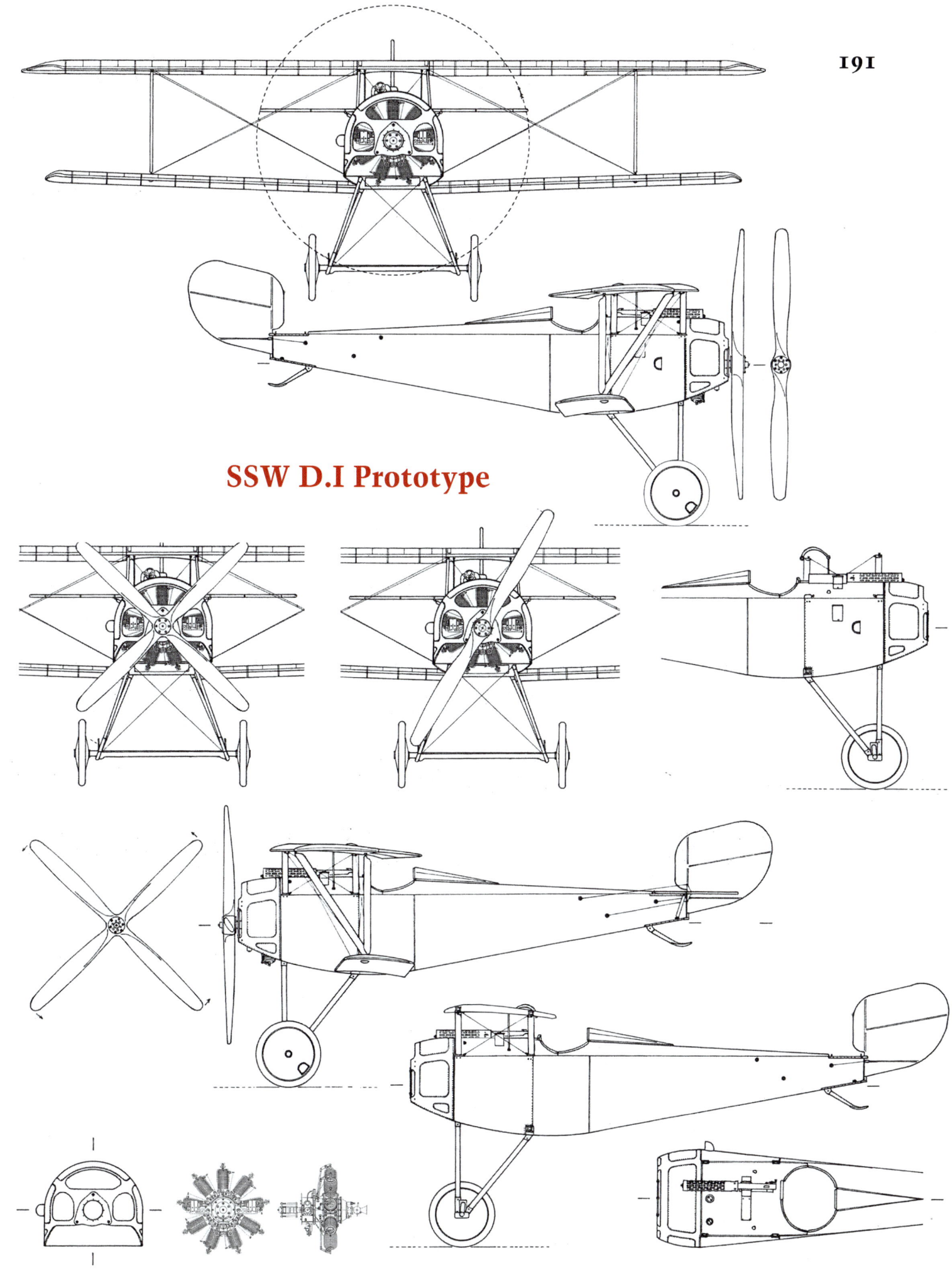

SSW D.I Prototype

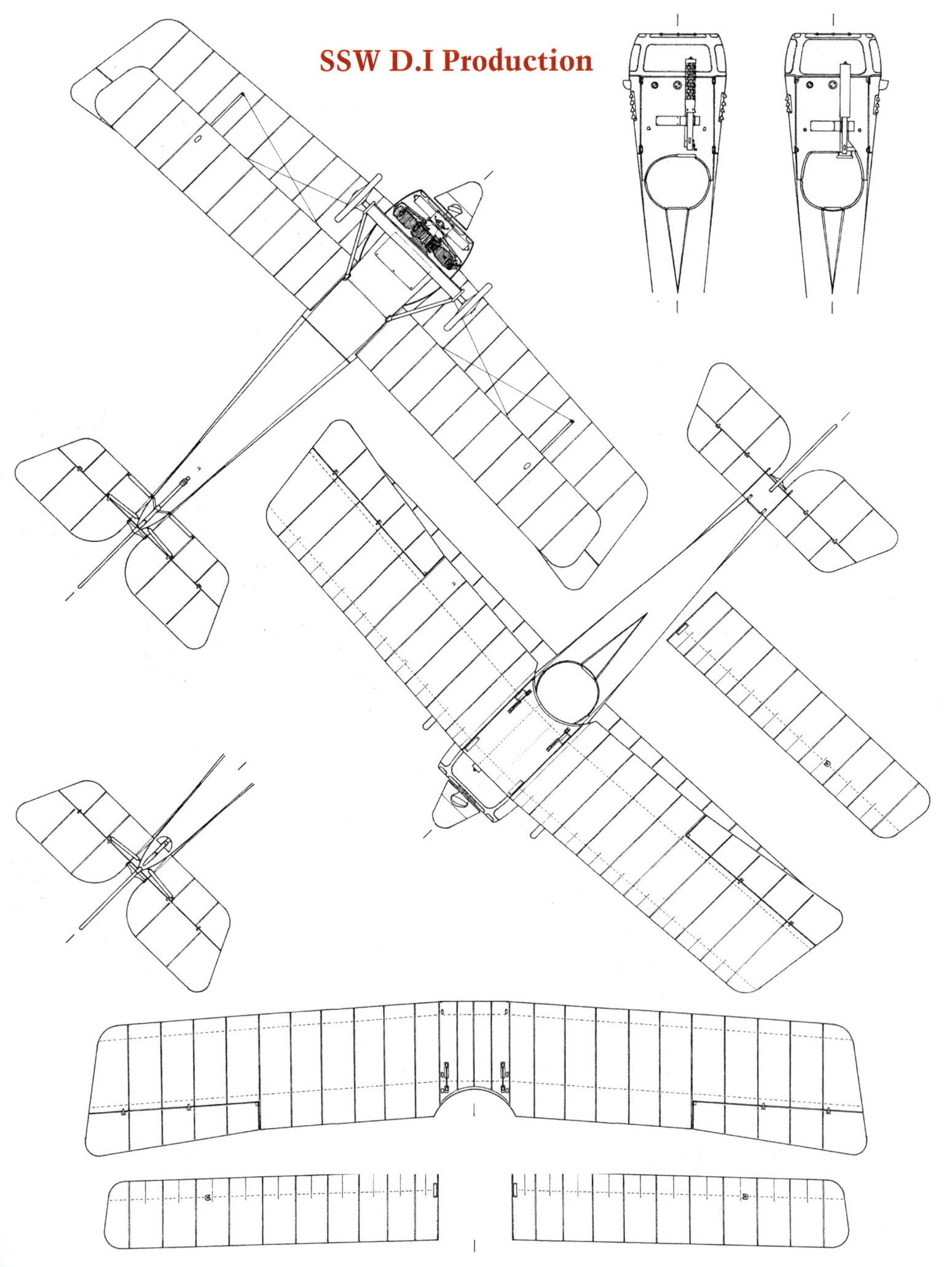

SSW D.I Production

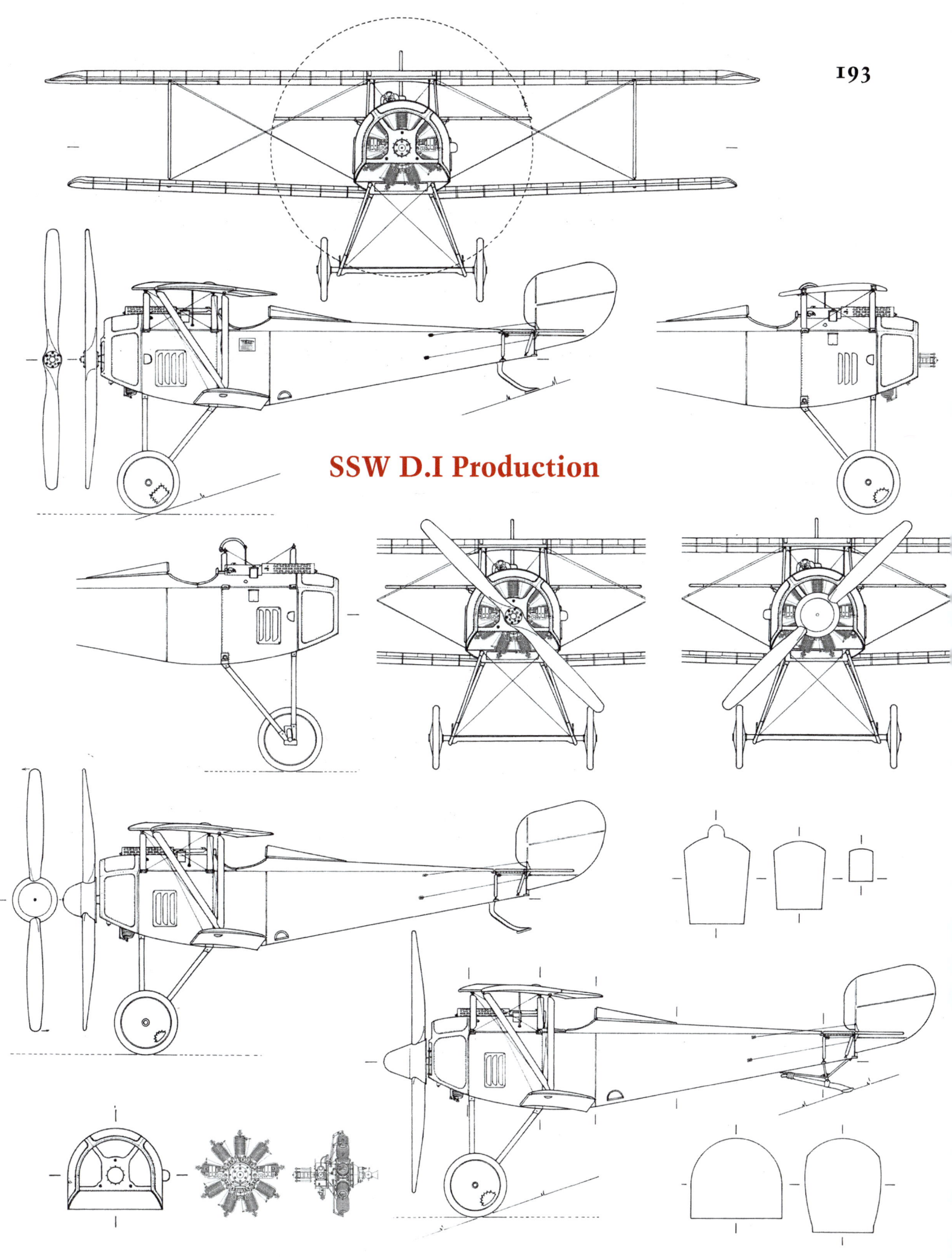

SSW D.I Production

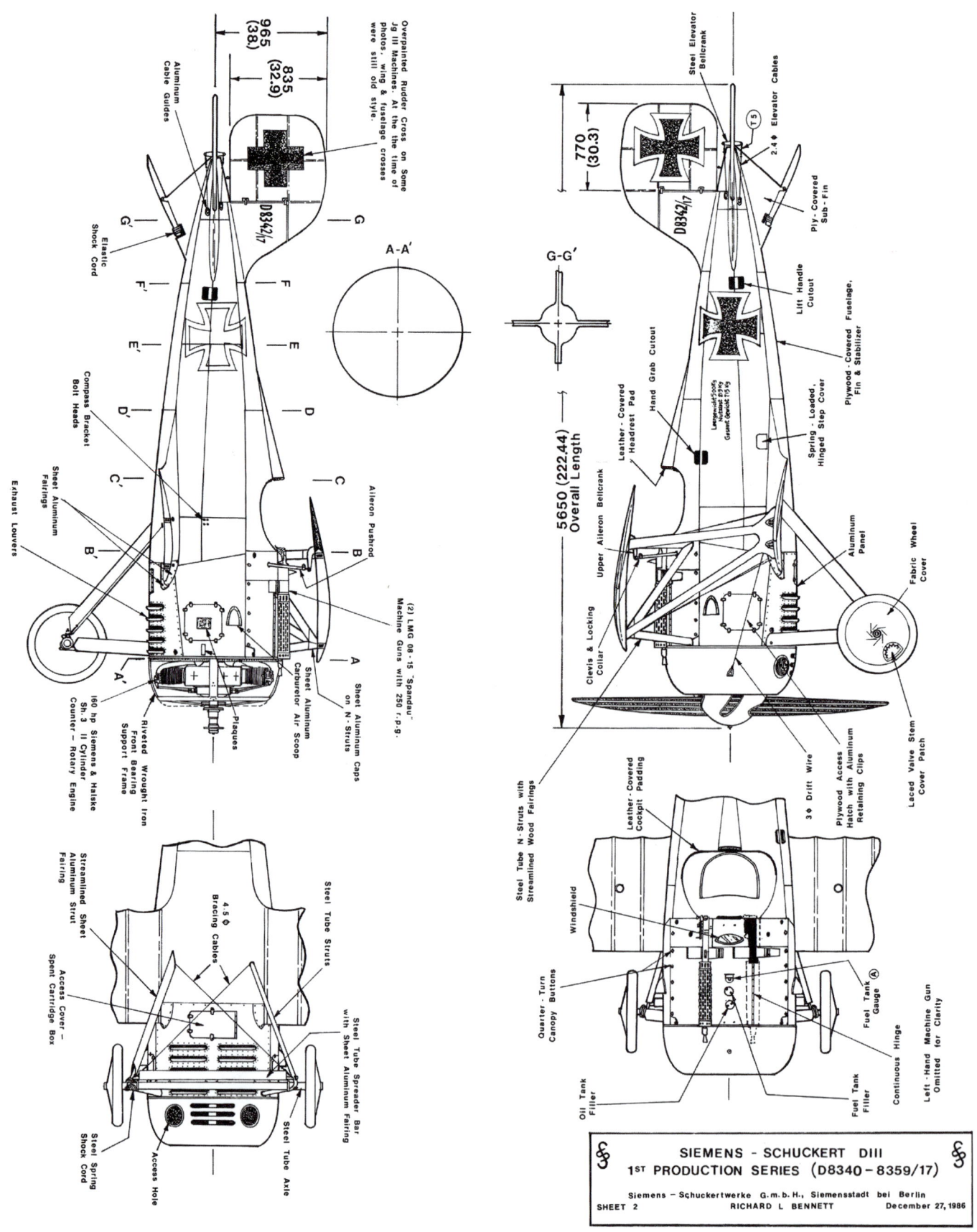

SIEMENS - SCHUCKERT DIII
1ST PRODUCTION SERIES (D8340 – 8359/17)

Siemens - Schuckertwerke G.m.b.H., Siemensstadt bei Berlin

SHEET 2   RICHARD L BENNETT   December 27, 1986

In March 1918, the first 20 SSW DIII's were shipped to the front for operational trials, D8340 through 8345 going to Jagdgeschwader III, and the rest, Jg II.

Although the DIII's drew high praise for their climb rate and maneuverability, numerous engine seizures led to their withdrawal from service. The surviving machines were returned to Siemens late May 1918 for refitting with new engines, cutaway cowlings, and revised wings and rudders.

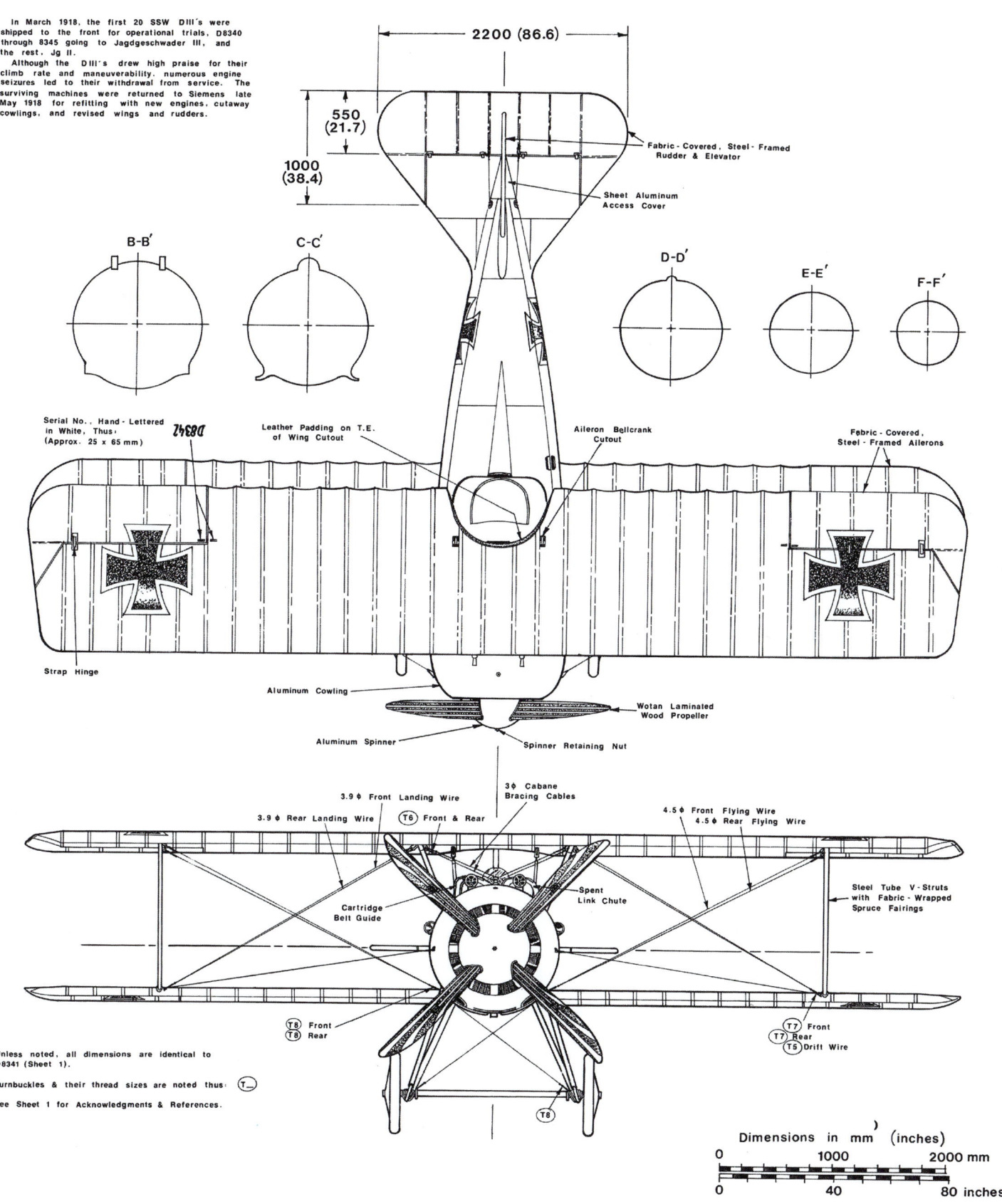

2200 (86.6)

550 (21.7)

1000 (38.4)

B-B'   C-C'   D-D'   E-E'   F-F'

Fabric - Covered, Steel - Framed Rudder & Elevator

Sheet Aluminum Access Cover

Serial No., Hand - Lettered in White, Thus: (Approx. 25 x 65 mm)

Leather Padding on T.E. of Wing Cutout

Aileron Bellcrank Cutout

Fabric - Covered, Steel - Framed Ailerons

Strap Hinge

Aluminum Cowling

Wotan Laminated Wood Propeller

Aluminum Spinner

Spinner Retaining Nut

3.9 φ Front Landing Wire

3.9 φ Rear Landing Wire

T6 Front & Rear

3 φ Cabane Bracing Cables

4.5 φ Front Flying Wire
4.5 φ Rear Flying Wire

Cartridge Belt Guide

Spent Link Chute

Steel Tube V - Struts with Fabric - Wrapped Spruce Fairings

T8 Front
T8 Rear

T7 Front
T7 Rear
T5 Drift Wire

Unless noted, all dimensions are identical to D8341 (Sheet 1).

Turnbuckles & their thread sizes are noted thus: T_

See Sheet 1 for Acknowledgments & References.

T8

Dimensions in mm (inches)

0   1000   2000 mm

0   40   80 inches

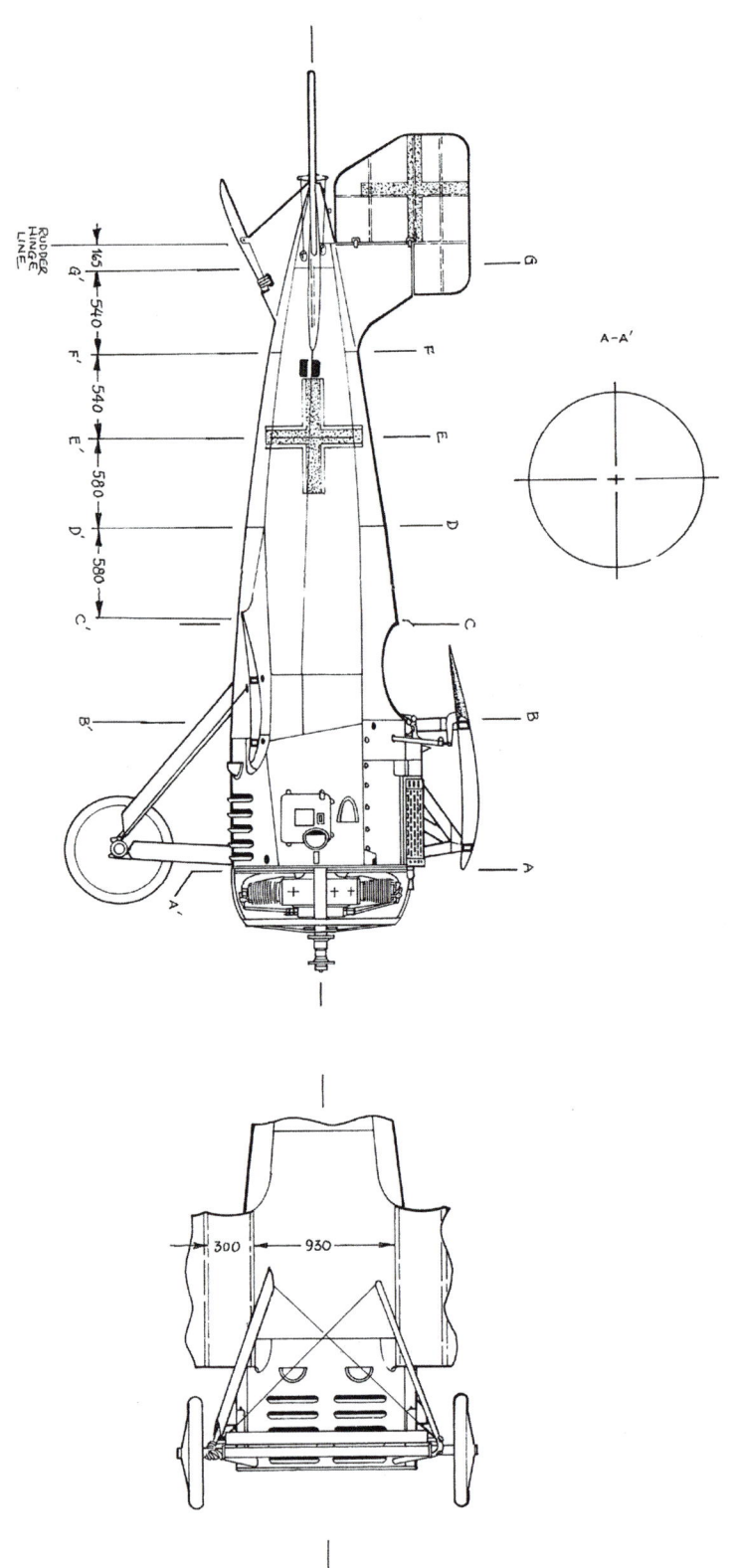

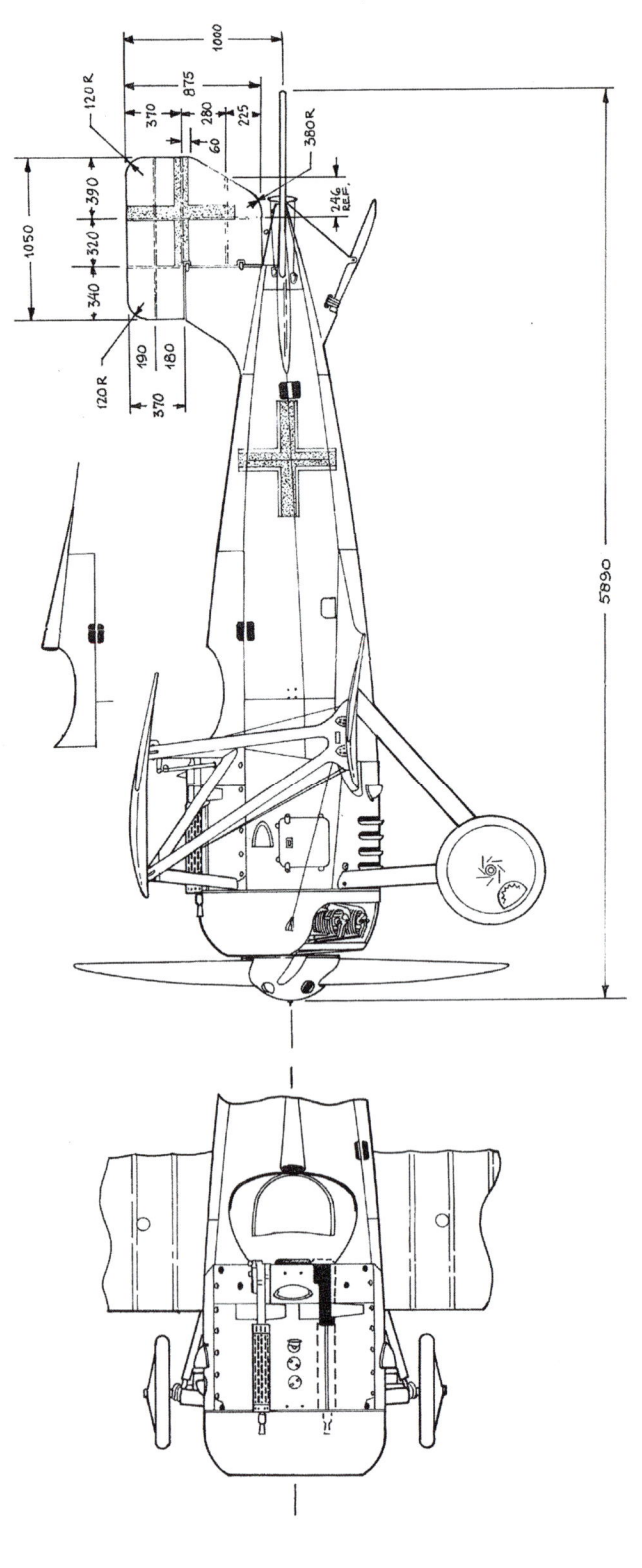

A-A'

SIEMENS - SCHUCKERT D.III
- SERIES 1 (D.8340 - 8359/17) REBUILT
- SERIES 2 (D.1600 - 1629/18) REBUILT
- SERIES 3 (D.3007 - 3026/18
           D.3037 - 3046/18)

Richard L. Sanders 10/6/99

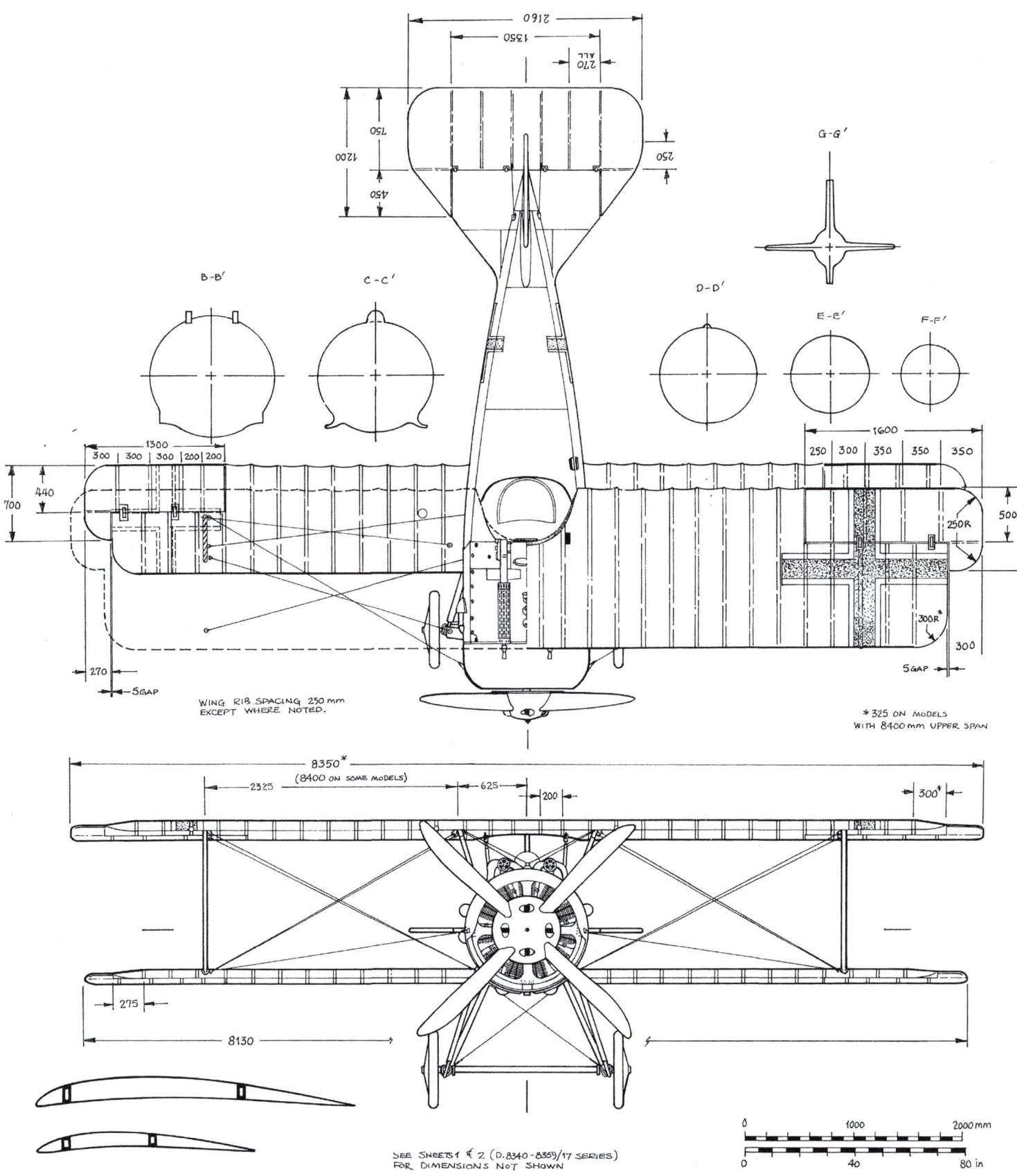

WING RIB SPACING 250 mm
EXCEPT WHERE NOTED.

* 325 ON MODELS
WITH 8400 mm UPPER SPAN

(8400 ON SOME MODELS)

SEE SHEETS 1 & 2 (D.8340-8359/17 SERIES)
FOR DIMENSIONS NOT SHOWN

0   1000   2000 mm

0   40   80 in

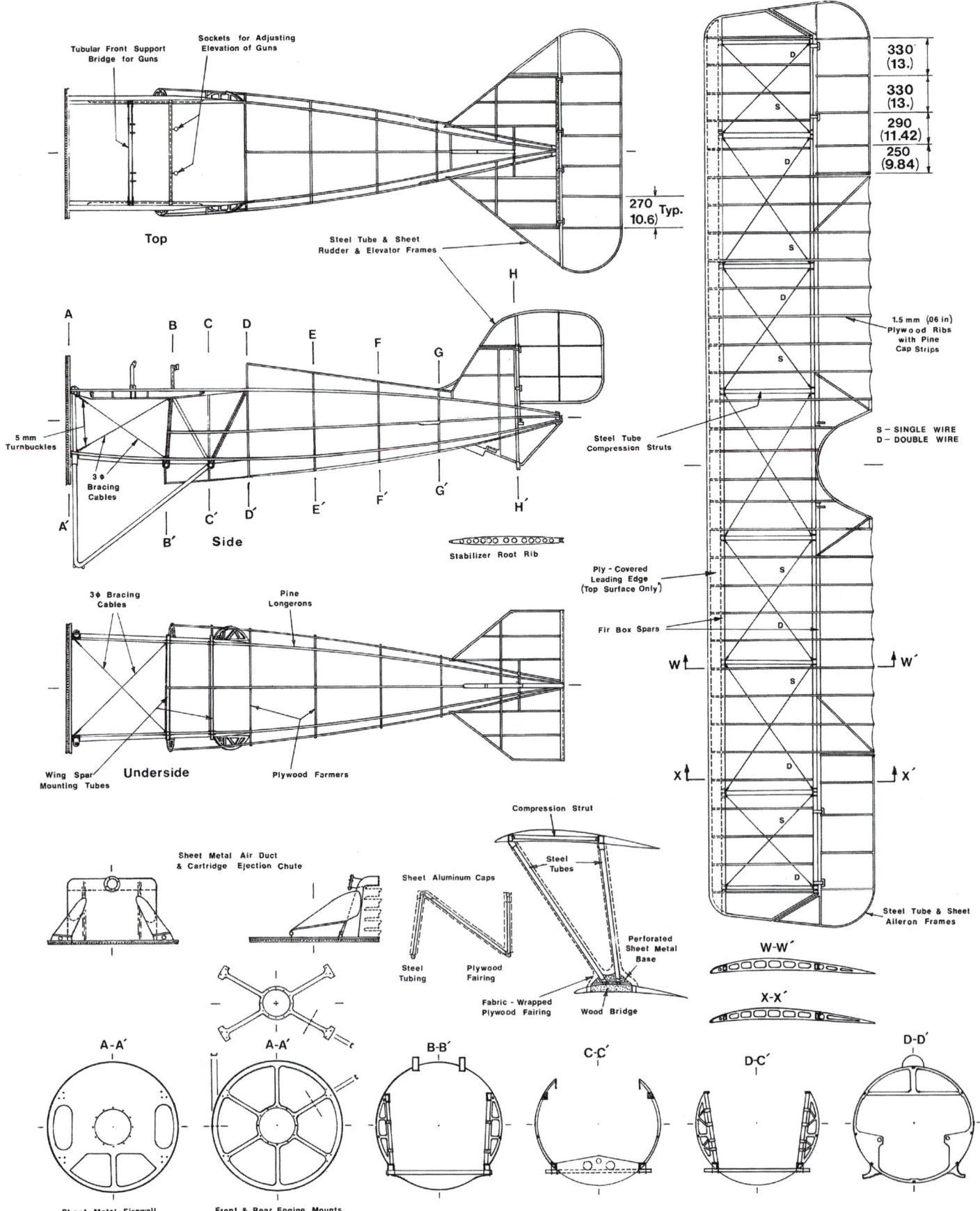

Tubular Front Support Bridge for Guns

Sockets for Adjusting Elevation of Guns

Top

Steel Tube & Sheet Rudder & Elevator Frames

270 (10.6) Typ.

A B C D E F G H

5 mm Turnbuckles

3 φ Bracing Cables

A' B' C' D' E' F' G' H'

Side

Stabilizer Root Rib

3 φ Bracing Cables

Pine Longerons

Underside

Wing Spar Mounting Tubes

Plywood Formers

330 (13.)
330 (13.)
290 (11.42)
250 (9.84)

1.5 mm (.06 in) Plywood Ribs with Pine Cap Strips

S — SINGLE WIRE
D — DOUBLE WIRE

Steel Tube Compression Struts

Ply - Covered Leading Edge (Top Surface Only)

Fir Box Spars

W   W'

X   X'

Steel Tube & Sheet Aileron Frames

Sheet Metal Air Duct & Cartridge Ejection Chute

Sheet Aluminum Caps

Compression Strut

Steel Tubes

Steel Tubing

Plywood Fairing

Perforated Sheet Metal Base

Fabric - Wrapped Plywood Fairing

Wood Bridge

W-W'

X-X'

A-A'
Sheet Metal Firewall

A-A'
Front & Rear Engine Mounts

B-B'

C-C'

D-C'

D-D'

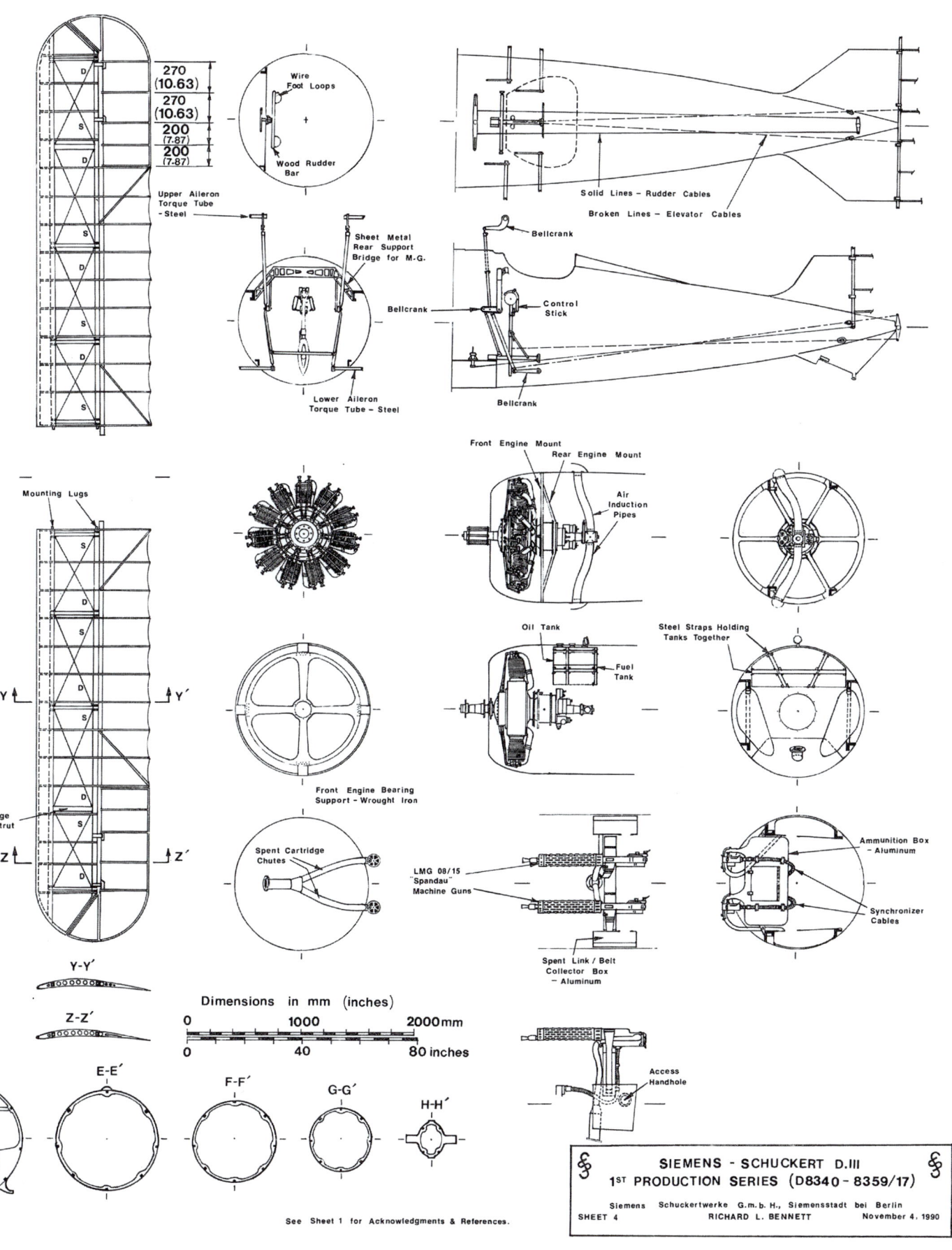

270
(10.63)

270
(10.63)

200
(7.87)

200
(7.87)

Wire
Foot Loops

Wood Rudder
Bar

Upper Aileron
Torque Tube
– Steel

Sheet Metal
Rear Support
Bridge for M-G.

Bellcrank

Bellcrank

Control
Stick

Bellcrank

Lower Aileron
Torque Tube – Steel

Solid Lines – Rudder Cables

Broken Lines – Elevator Cables

Mounting Lugs

Y   Y′

dge
Strut

Z   Z′

Front Engine Mount

Rear Engine Mount

Air
Induction
Pipes

Oil Tank

Fuel
Tank

Steel Straps Holding
Tanks Together

Front Engine Bearing
Support – Wrought Iron

Spent Cartridge
Chutes

LMG 08/15
"Spandau"
Machine Guns

Spent Link / Belt
Collector Box
– Aluminum

Ammunition Box
– Aluminum

Synchronizer
Cables

Access
Handhole

Y-Y′

Z-Z′

Dimensions in mm (inches)

0   1000   2000 mm

0   40   80 inches

E-E′   F-F′   G-G′   H-H′

See Sheet 1 for Acknowledgments & References.

SIEMENS - SCHUCKERT D.III
1ST PRODUCTION SERIES (D8340 - 8359/17)
Siemens Schuckertwerke G.m.b.H., Siemensstadt bei Berlin
SHEET 4   RICHARD L. BENNETT   November 4, 1990

SSW D.IV

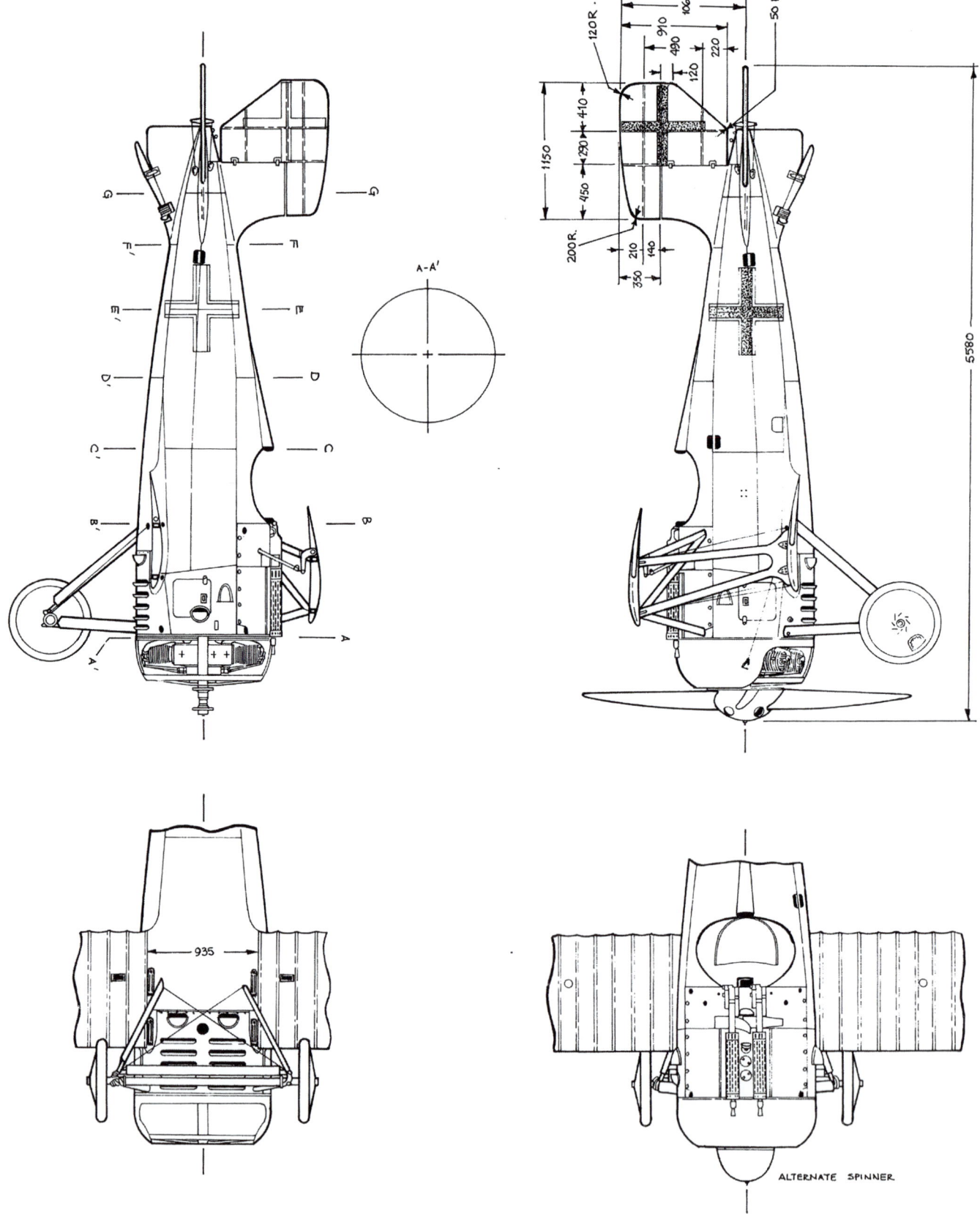

ALTERNATE SPINNER

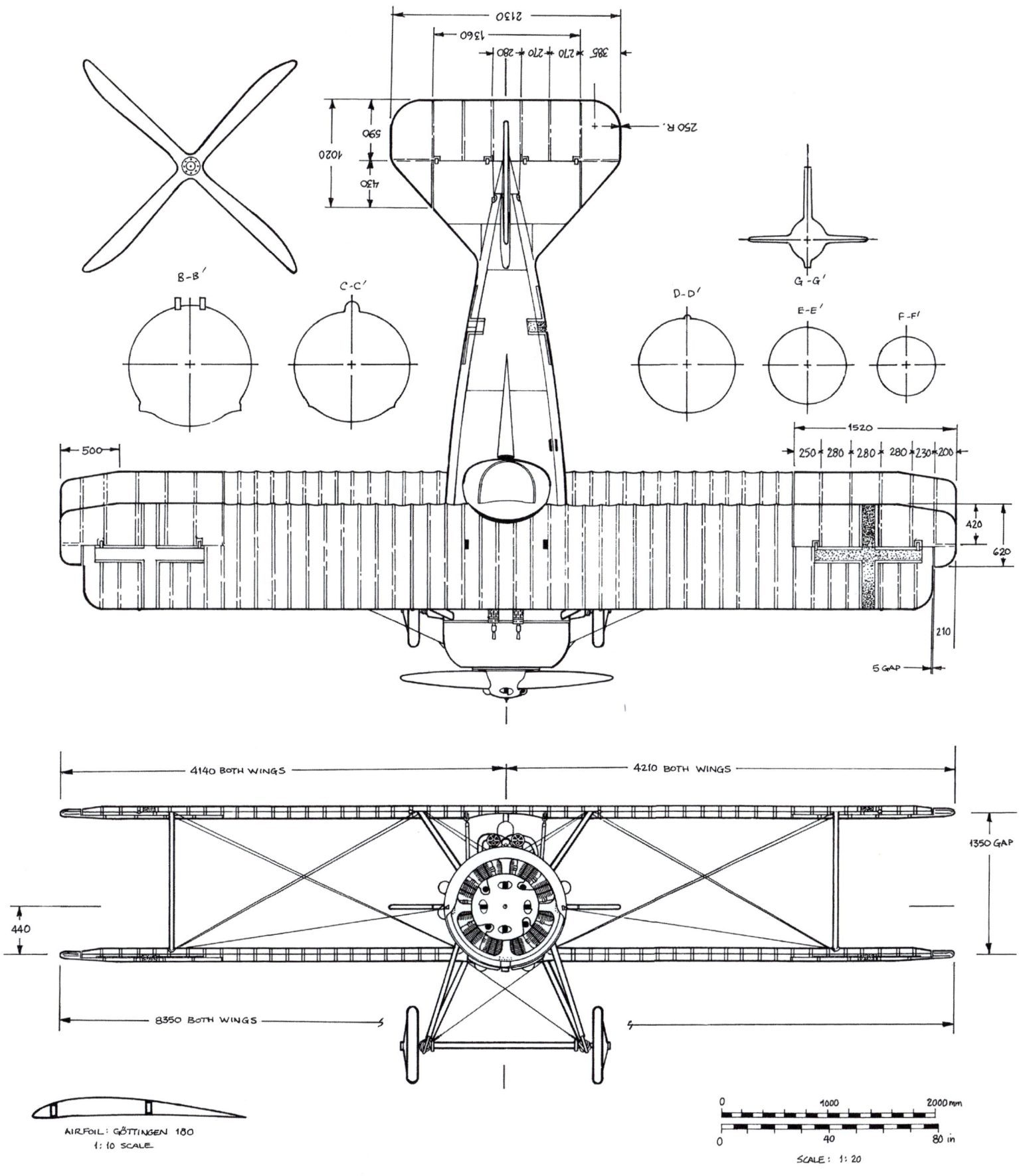

AIRFOIL: GÖTTINGEN 180
1:10 SCALE

SCALE: 1:20

SSW D.VI

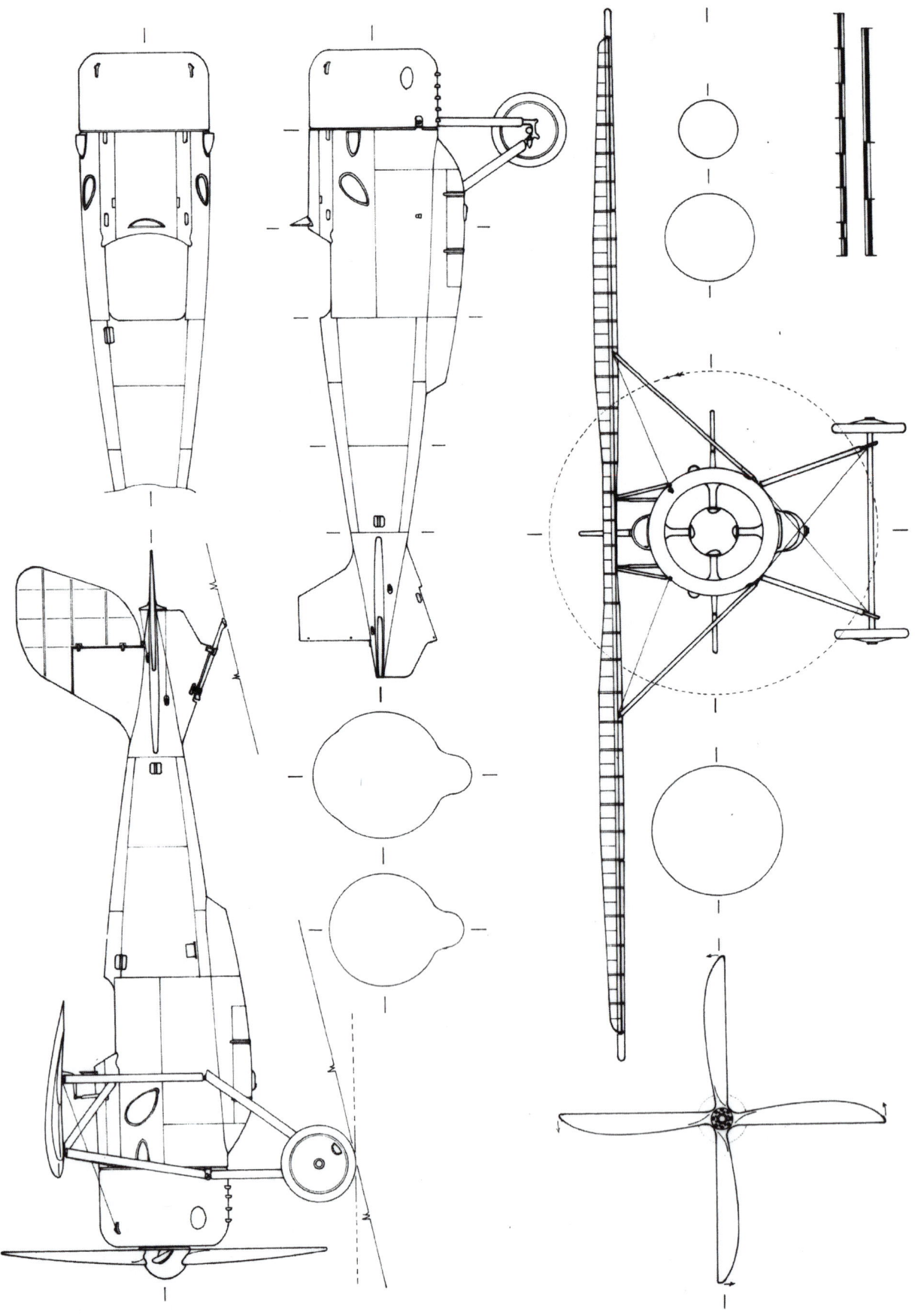

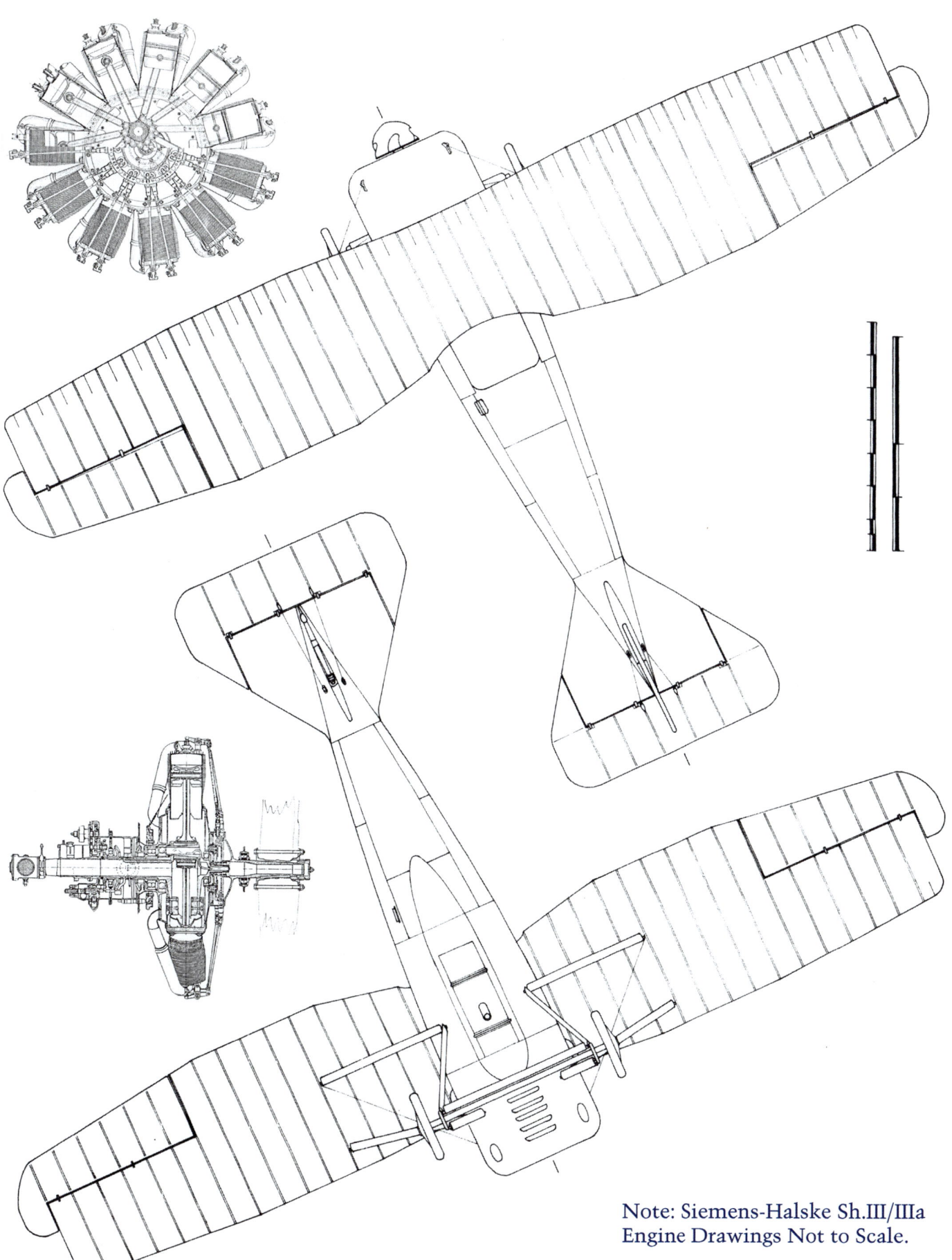

Note: Siemens-Halske Sh.III/IIIa
Engine Drawings Not to Scale.

SSW L.I

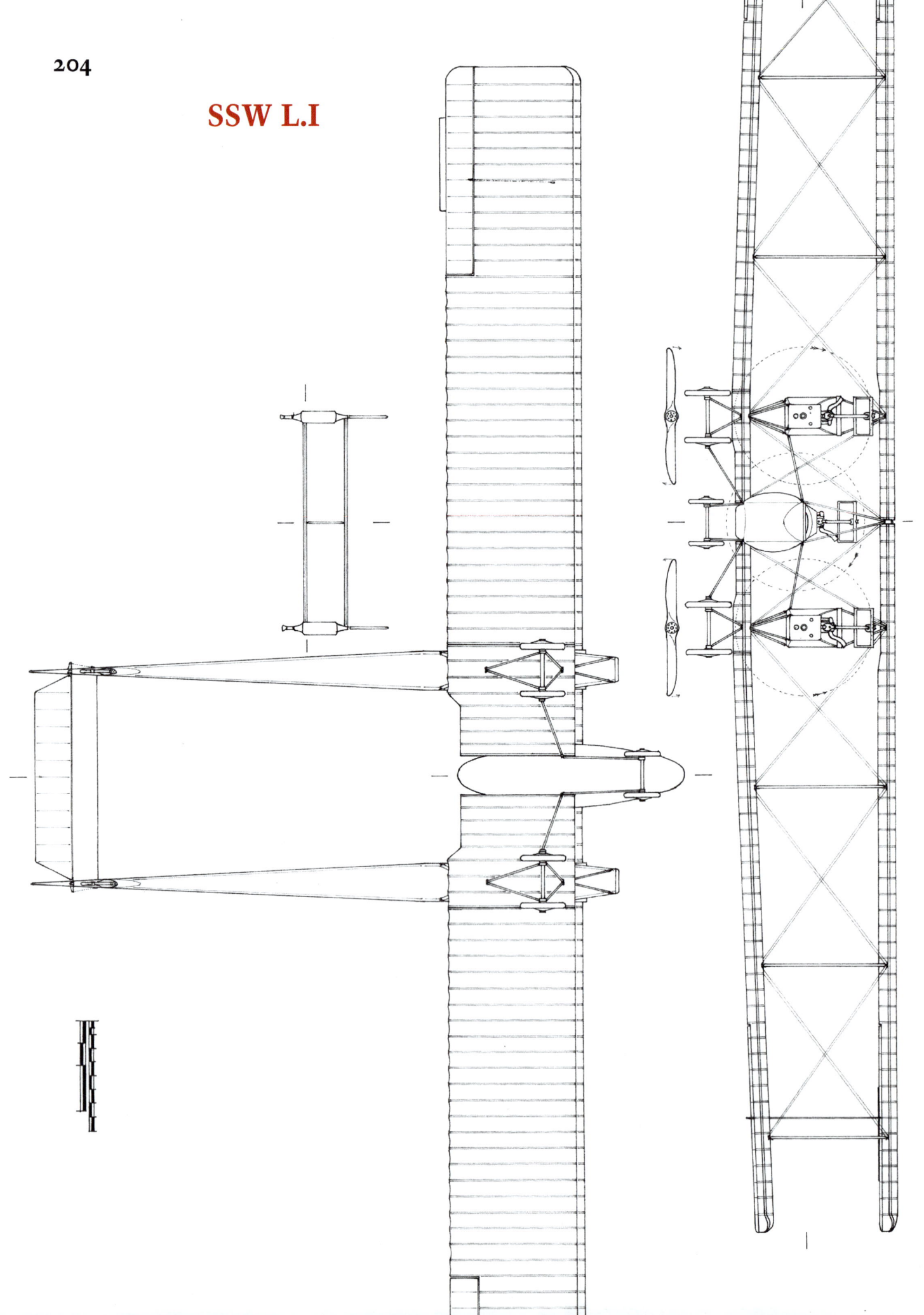

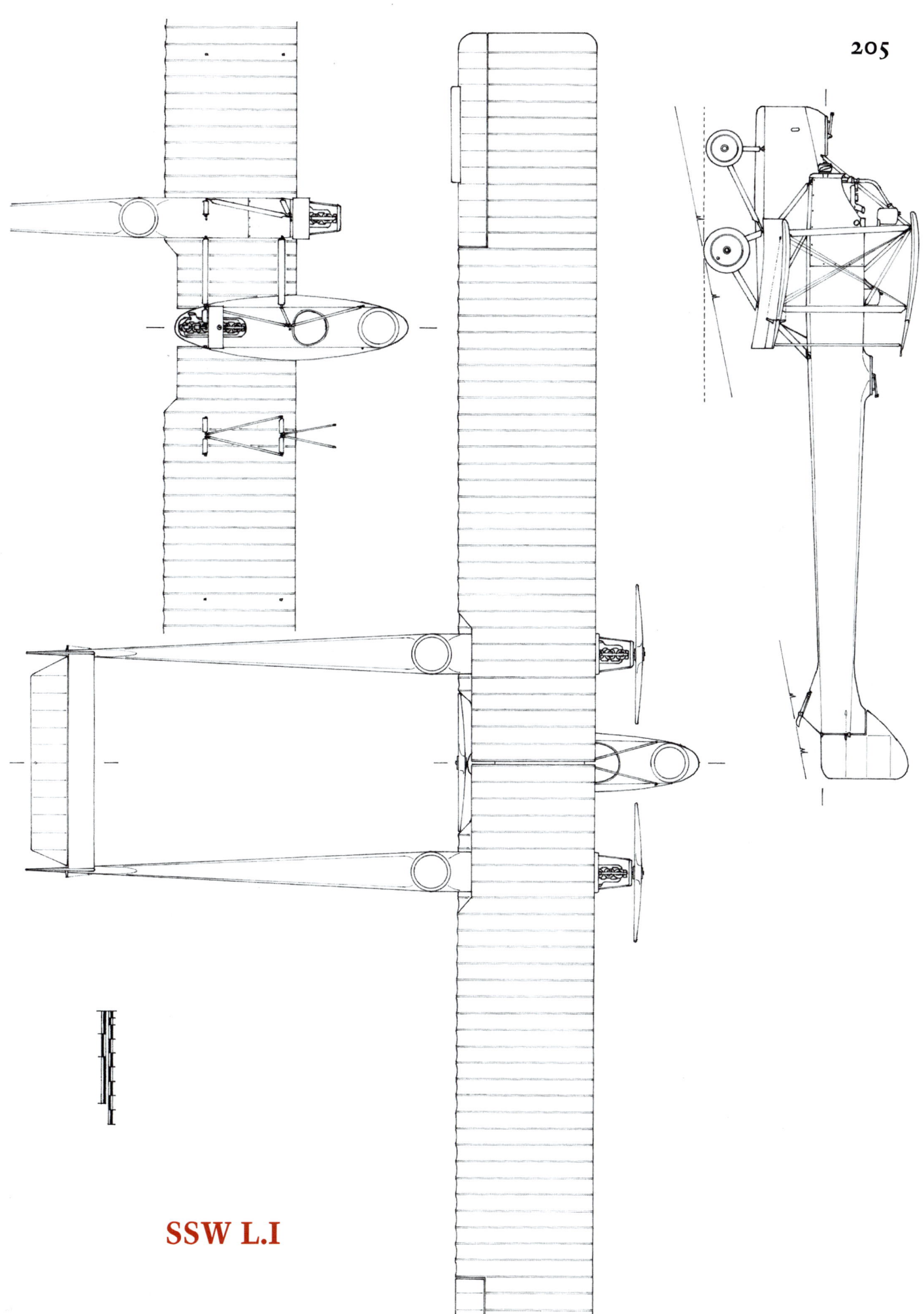

SSW L.I

Afterword

Above: *Lt.* Ernst Udet (*Jasta* 4, at left) and members of his ground crew in front of his new SSW D.III at Metz. Udet scored 62 victories, second only to the Red Baron among German aces, and was awarded the *Pour le Mérite*. Surviving the war, Udet became a famous stunt pilot and movie pilot. Udet was presented with this SSW D.III late in the war, after scoring all his victories. It is unknown if he flew this aircraft in combat, but he did not score any victories in it.

The number two German ace, Ernst Udet survived the war and became famous during the postwar years as a movie and stunt pilot.

Udet achieved his final two (of 62) victories on 26 September, but was slightly wounded in the thigh in the same dogfight. He left for a brief recuperative leave, returning to the *Geschwader* on October 3, when it was based at Metz (Metz-Frescaty airfield). On 6 October Friedrich Noltenius wrote that: "On this date a French balloon stood for the first time. I prepared for an attack against it. But prior to it I had a little mock fight with Udet who flew the Siemens D.III with the powerful Remag (sic) engine that Udet had brought with him upon returning from leave after being wounded in September. It was impossible to match the performance of this combination." And, Noltenius was flying a BMW-engined Fokker D.VII. During his week back at the front, Udet also had his Fokker D.VII avalable to fly.

On 9 October, *JG* I moved to Marville, and on 11 October Udet was transferred to a post with *Idflieg*. From van Ishovern's biography of Udet: "(Udet) was assigned to *Flieger Ersatz Abteilung* 3 – a reserve unit at Gotha. In this capacity, from 15 to 18 October, he was ordered to visit the Rhemag-Rhenania engine factory, which was building the Siemens Sh.III rotary engine under license. His fiancee, Lo, joined him in Mannheim for the three days. His next job was to attend the third and final fighter evaluation trials at Adlershof. There were 15 types being tested. The trials ended on 31 October with a communal dinner at the Bristol Hotel. The general mood was sombre…"

Pressured in the mid-1930s to join the new *Luftwaffe*, he became chief of the technical department and a *Generaloberst*. Unsuited for a technical job and unable to cope with the political infighting, he commited suicide Nov. 17, 1941.

Above: *Lt*. Ernst Udet in his new SSW D.III at Metz. Eleonore Zink was Udet's fiancee, and he had her 'pet' name painted on the fuselage of some of his aircraft, including this D.III, as a personal marking.

Below: *Lt*. Ernst Udet in front of his new SSW D.III, almost certainly 8350/17, at Metz-Frescaty airfield. The man in the center with the mustache appears to be *Ingenieur* Kaendler from SSW.

This Page: More views of *Lt.* Ernst Udet and his SSW D.III at Metz before and after "LO!", his fiancee's pet name, was painted on the fuselage.

Printed in Great Britain
by Amazon.co.uk, Ltd.,
Marston Gate.